Regulatory T Cells - New Insights

Edited by Xuehui He

Published in London, United Kingdom

Regulatory T Cells - New Insights
http://dx.doi.org/10.5772/intechopen.98075
Edited by Xuehui He

Contributors
Ceren Ciraci, Ece Oylumlu, Goksu Uzel, Lubeyne Durmus, Meric Tas, Damla Gunes, Chao Gu, SangKon Oh, Yang Liu, Tiezheng Hou, Huiqin Hao, Rahmad Aji Prasetya, Devyani Diah Wulansari

First published in London, United Kingdom, 2023 by IntechOpen
IntechOpen is the global imprint of INTECHOPEN LIMITED, registered in England and Wales, registration number: 11086078, 5 Princes Gate Court, London, SW7 2QJ, United Kingdom

British Library Cataloguing-in-Publication Data
A catalogue record for this book is available from the British Library

Additional hard and PDF copies can be obtained from orders@intechopen.com

Regulatory T Cells - New Insights
Edited by Xuehui He
p. cm.
Print ISBN 978-1-80356-056-4
Online ISBN 978-1-80356-057-1
eBook (PDF) ISBN 978-1-80356-058-8

Meet the editor

Dr. Xuehui He received her Ph.D. at Tübingen University, Germany. After graduation, she moved to Radboud University Medical Center, Netherlands, to investigate the regulation of the human immune system in relation to immune-mediated-diseases. Her research interests include the maintenance of human Treg stability and identity, developing novel small-molecule-inhibitors, and immune profiling of patients with distinct autoimmune diseases including psoriasis, psoriatic arthritis, and type 1 diabetes as well as cancer patients undergoing treatment with checkpoint inhibitors. With the advent of high-throughput techniques and the availability of multi-omics data generated from a large set of samples, Dr. Xuehui He is also specialized in applying various tools and methods for data integration, analysis and interpretation. She believes that combining multi-omics data to highlight the interrelationships of the involved biomolecules and their functions is crucial to fully understand human health and diseases.

Contents

Preface

The immune system defends the body from invading microbes and protects against disease. When pathogens like bacteria or viruses pass through the skin or mucosal membranes the innate immune cells are initially triggered, inducing a quick but non-specific response. Thereafter, the adaptive immune system is activated, which is more powerful but takes a couple of days to be fully activated. The main carriers of the adaptive immune system are lymphocytes, also called white blood cells, which include B cells and T cells. Antibody response is mainly mediated by B cells, whereas T cells are involved in cell-mediated immune response. Both CD4 and CD8 T cells can be further grouped into distinct subsets based on their differentiated cell functions. Effector cell subsets induce a proinflammatory immune response, thus delaminating pathogens and antigens, while regulatory cell subsets induce an anti-inflammatory immune response that is crucial for immune homeostasis and preventing autoimmune disease. Various immune cells have been found to have a regulatory capacity, such as regulatory T cells (Treg), regulatory B cells (Breg), macrophages, myeloid-derived suppressor cells (MDSCs), and tolerogenic dendritic cells (tol-DCs, tDCs). This book focuses on regulatory T cells, including CD4+CD25+Foxp3 Treg cells and interleukin 10 (IL-10)-producing type 1 regulatory T (Tr1) cells.

Tr1 cells are currently identified by the unique cytokine profile of high levels of IL-10 and transforming growth factor beta (TGFb), low levels of IL-2, and the absence of IL-4 after stimulation [1]. The specific cell surface markers used for identification of Tr1 cells include CD49b and LAG-3. Tr1 cells do not constitutively express Foxp3. Foxp3 is required neither for Tr1 cell induction nor suppressive function since it can be isolated from peripheral blood of patients with immune dysregulation polyendocrinopathy enteropathy X-linked (IPEX) syndrome, a disease condition due to *FOXP3* gene mutations [2]. Tr1 cells are multifaceted suppressors that use several immune regulation mechanisms to achieve tolerance. Tr1 cells can inhibit T-cell responses via cell-cell contact, metabolic disruption, and cytolysis. Chapter 1 discusses the potential applications of Tr1 cells in cell-based therapy.

Sakaguchi introduced the concept of T-cell-mediated suppression by describing a subset of CD4+ T cells with co-expression of interleukin-2 receptor (IL-2R) alpha-chain (CD25), which were found to be anergic and to suppress autoimmune disease in thymecotmized mice [3]. Later, Foxp3 was identified as a master regulator gene for Treg development and function. CD4+CD25highCD127low T cells in human peripheral blood can identify a population with high Foxp3 expression. Several other markers were also described for the identification of Treg cells in humans, such as CTLA-4, GITR, TIGIT, and CD49d [4, 5]. To keep the Treg cell lineage stability and identity, epigenetic mechanisms including DNA methylation histone modifications and nucleosome positioning are essential for controlling Treg signature gene expression and thus maintaining suppressive function [6]. It is known that Foxp3 plays a critical role as a transcriptional activator and repressor. Foxp3 can bind to more than 700 genes involved in various cellular programs including TCR signaling, cell communication, and transcriptional regulation [7]. Due to Treg cells' crucial role in immune homeostasis and tolerance, therapeutic strategies for autoimmune disease and transplantation patients have been developed that target Treg cells directly or indirectly. Chapters 2–4 present the state of the art in therapeutic interventions of Treg cells in immune regulation, application of immune checkpoint inhibitors in cancer treatment, and pattern recognition receptor-mediated regulatory T-cell functions.

The potential therapeutic use of Treg cells for the treatment of autoimmune diseases, transplant rejection, and graft-versus-host disease is currently being investigated in clinical trials. Treg cells might also help to control the autoinflammatory adverse events that are often associated with checkpoint inhibitor treatment for cancer patients. However, engineering Treg cells to enhance specificity, stability, functional activity and delivery, and efficiency is challenging. We anticipate that the new multi-omics tools and CRISPR-Cas9-mediated technology will allow for the rapid growth of Treg-based immunotherapies.

Xuehui He
Department of Laboratory Medicine,
Lab Medical Immunology,
Radboud University Medical Centre,
Nijmegen, The Netherlands

References

[1] Groux H. Type 1 T-regulatory cells: Their role in the control of immune responses. Transplantation. 2003; **75**(9 Suppl):8S-12S

[2] Passerini L, Allan SE, Battaglia M, Di Nunzio S, Alstad AN, Levings MK, et al. STAT5-signaling cytokines regulate the expression of FOXP3 in CD4+CD25+ regulatory T cells and CD4+CD25- effector T cells. International Immunology. 2008;**20**(3):421-31

[3] Setoguchi R, Hori S, Takahashi T, Sakaguchi S. Homeostatic maintenance of natural Foxp3(+) CD25(+) CD4(+) regulatory T cells by interleukin (IL)-2 and induction of autoimmune disease by IL-2 neutralization. The Journal of Experimental Medicine. 2005;**201**(5):723-35

[4] Gagliani N, Magnani CF, Huber S, Gianolini ME, Pala M, Licona-Limon P, et al. Coexpression of CD49b and LAG-3 identifies human and mouse T regulatory type 1 cells. Nature Medicine. 2013;**19**(6):739-46

[5] He X, Landman S, Bauland SC, van den Dolder J, Koenen HJ, Joosten I. A TNFR2-Agonist facilitates high purity expansion of human low purity Treg cells. PloS One. 2016;**11**(5):e0156311

[6] Scheinecker C, Goschl L, Bonelli M. Treg cells in health and autoimmune diseases: New insights from single cell analysis. Journal of Autoimmunity. 2020;**110**:102376

[7] Marson A, Kretschmer K, Frampton GM, Jacobsen ES, Polansky JK, MacIsaac KD, et al. Foxp3 occupancy and regulation of key target genes during T-cell stimulation. Nature. 2007;**445**(7130):931-935

Chapter 1

Type 1 Regulatory T Cells and Their Application in Cell Therapy

Chao Gu and SangKon Oh

Abstract

Critical roles of regulatory T cells (Tregs) in the maintenance of immune homeostasis by controlling unwanted types of immune responses have been well documented. Therefore, Treg-based therapeutic strategies for inflammatory diseases have long been investigated. Type 1 regulatory T (Tr1) cells and $Foxp3^+$ Tregs are two major subsets of regulatory $CD4^+$ T cells. In contrast to $Foxp3^+$ Tregs, the master transcription regulator for Tr1 cells still remains elusive. Nevertheless, Tr1 cells are generally defined as a specialized subset of $CD4^+$ T cells, which are induced in the periphery during antigen exposure in tolerogenic condition. As one of their key features, Tr1 cells express immunosuppressive cytokine IL-10, which can repress the function of effector immune cells independently of Foxp3 expression. In this book chapter, we discuss the recent developments in the field of Tr1 cells, including major characteristics of Tr1 cells, methods for Tr1 induction as well as their therapeutic potentials in immune-mediated diseases.

Keywords: Tr1 cells, Tregs, IL-10, Foxp3, CD49b, LAG-3, cell therapy, immune regulation

1. Introduction

The immune system is a delicate network consisting of a variety of cellular and molecular components that are designed to protect the host by clearing invading foreign pathogens as well as altered self antigens [1]. In addition, immune system is also equipped with a fine-tuned regulatory machinery that can maintain the balance between activation and suppression of immune responses to achieve immune homeostasis and tolerance.

Studies on the mechanisms of immune regulation have revealed multiple different cell types, including subsets of T cells [2, 3], B cells [4, 5], NK cells [6, 7], and myeloid-derived suppressor cells (MDSCs) [8, 9], with immune regulatory function. Among them, two types of regulatory $CD4^+$ T cells, namely type 1 regulatory T (Tr1) cells and Forkhead box protein $P3^+$ regulatory T cells ($Foxp3^+$ Tregs) are best studied so far. These two types of Tregs have, to some extent, overlapping functions in immune regulation. For example, they both can downregulate unwanted types of immune responses and play important roles in the maintenance of immune tolerance in general. However, mounting evidence suggests that Tr1 cells and $Foxp3^+$ Tregs are distinct populations of regulatory $CD4^+$ T cells [10] and more importantly, they can also display different immune regulatory properties [11–13]. For example, human Tr1 cells, but not $Foxp3^+$ Tregs, have been reported to secrete IL-22 and protect gut epithelial cells from TNFα-induced damage [14, 15].

Foxp3 is known to be the master transcription regulator for Foxp3$^+$ Tregs [16–18]. Both naïve and memory CD4$^+$ T cells are known to differentiate into Tregs with induced Foxp3 expression [19]. Mutation in the *FOXP3/Foxp3* gene gives rise to hyperactive T cell responses [16, 20]. In contrast, though many transcription factors have been reported to transactivate *IL10/Il10* gene, which plays an important role in the differentiation of Tr1 cells, the master transcription regulator for Tr1 cells is still under investigation. Therefore, several key features have been proposed to identify Tr1 cells. First, Tr1 cells produce predominantly IL-10 and TGFβ. However, Tr1 cells can also express different amounts of other cytokines, including IFNγ [21, 22], depending on the microenvironments where Tr1 cells localize. Second, Tr1 cells exhibit suppressive functions without constitutive Foxp3 expression. IL-10 expressed by Tr1 cells plays a major role in Tr1 cell-mediated immune suppression. Third, CD49b and Lymphocyte-activation gene 3 (LAG-3) have been proposed as surface markers for both human and mouse Tr1 cells [10]. However, whether they contribute to the immunosuppressive functions of Tr1 cells remains elusive so far. In addition, CD49b and LAG-3 are also expressed by other cell types, including CD8$^+$ T cells and B cells, which can also express IL-10. In addition to CD49b and LAG-3, Tr1 cells can express other surface proteins, including CTLA-4, PD-1, T cell immunoglobulin and mucin domain-containing protein 3 (TIM-3), T cell immunoreceptor with Ig and ITIM domains (TIGIT), CD226, inducible T cell costimulator (ICOS), and CCR5. Of note, these surface molecules can be expressed not only by Tr1 cells, but also by Foxp3$^+$ Tregs and non-Treg populations, depending on the immune context [23].

Herein, we review the major characteristics of Tr1 cells and different experimental methods to induce Tr1 cells both *in vitro* and *in vivo*. We also summarize animal models and human diseases in which Tr1 cells are indispensable in controlling inflammatory immune responses. In addition, we recapitulate clinical trials using Tr1 cells as immunotherapeutics. Lastly, we discuss the future perspectives and major questions to be addressed in the field of Tr1 cells.

2. Tr1 cells and Foxp3$^+$ Tregs are distinct populations of regulatory T cells

Chronic stimulation of naïve CD4$^+$ T cells from both human and mouse in the presence of IL-10 has been reported to induce IL-10-producing antigen-specific immunosuppressive T cells *in vitro* [24]. More importantly, this induced CD4$^+$ T cell population can prevent the development of colitis *in vivo* induced by pathogenic CD4$^+$CD45RB^{hi} splenic T cells in mice [24]. Based on these findings, such antigen-specific IL-10-producing immunoregulatory CD4$^+$ T cells are therefore designated Tr1 cells [24]. Of note, the term "Tr1 cells" was coined several years before the initial studies reporting Foxp3$^+$ Tregs [16, 17], therefore, the question whether Tr1 cells and Foxp3$^+$ Tregs could be distinct populations of regulatory T cells remained unsolved at that time.

By utilizing the IL-10 reporter mice, in which the cellular source of IL-10 can be detected, the presence of Tr1 cells is further investigated *in vivo* [25, 26]. In a steady state, Tr1 cells have been found in multiple tissues of mice, including small intestine and spleen [10, 26]. More importantly, in the IL-10 and Foxp3 dual-reporter mice, the presence of Foxp3$^-$IL-10$^+$ Tr1 cells and Foxp3$^+$IL-10$^{+/-}$ Tregs, as well as their different distributions in tissues and developmental origins under steady state condition have been observed *in vivo*, suggesting that Tr1 cells and Foxp3$^+$ Tregs are different subsets [10, 26]. Of note, there is only low and transient Foxp3 expression upon activation of Tr1 cells, and functional Tr1 cells have been generated *in vitro* from Foxp3-mutated

CD4⁺ T cells of immunodysregulation polyendocrinopathy enteropathy X-linked syndrome (IPEX) patients [13], indicating that in contrast to Foxp3⁺ Tregs [16, 17], Tr1 cells do not require Foxp3 for their development or suppressive function [13].

Taken together, all these findings have identified that immunosuppressive Tr1 cells are different T cell population from Foxp3⁺ Tregs. More importantly, Tr1 cells also play important roles in the induction and maintenance of immune homeostasis.

3. Experimental induction of Tr1 cells in animal models *in vivo* and in humans *in vitro*

So far, multiple methods have been reported to induce Tr1 cells (**Table 1**). Different microbial components have been shown to induce and further promote Tr1

Reagent and Method	Mouse	References	Human	References
Heat-killed *Mycobacterium vaccae* suspension	+	[27]		
Bordetella pertussis filamentous hemagglutinin	+	[28]		
Cholera toxin	+	[29, 30]		
V. filiformis lysate	+	[31]		
Lactobacillus pentosus KF340	+	[32]		
B. breve	+	[33]		
Microbiota-derived short-chain fatty acids	+	[34]		
PEGylated G-CSF	+	[35]		
Bowman-Birk inhibitor	+	[36, 37]		
Rapamycin + IL-10	+	[38]		
Rapamycin + G-CSF	+	[39, 40]		
Vitamin D3 + dexamethasone	+	[41]	+	[41]
IL-10	+	[24]	+	[24]
TGFβ	+	[26, 42]		
IL-27	+	[43–49]	?	[50, 51]
IFNα			+	[52, 53]
IL-6			+	[54]
DC-ASGPR agonist			+	[22, 55]
ICOS–ICOSL ligation	+	[56]	+	[57]
Indoleamine-2,3-dioxygenase (IDO)			+	[58]
Activin-A	+	[59]	+	[60]
Retinoic acid (RA)	+	[61]	+	[62]
CD2–CD58 ligation			+	[63]
Co-stimulation of CD46			+	[64]
Co-stimulation of CD55			+	[65]
Artificial APCs			+	[66, 67]

Table 1.
Factors that promote the generation of Tr1 cells in mice and humans.

cell differentiation in mice *in vivo*. For example, subcutaneous injection of heat-killed *Mycobacterium vaccae* suspension induces allergen-specific IL-10-producing Tr1-like cells that can protect mice against airway inflammation [27]; treatment with filamentous hemagglutinin from *Bordatella pertussis* inhibits IL-12, but induces IL-10 expression by dendritic cells (DCs), which in turn direct naïve CD4$^+$ T cell differentiation into Tr1 cells in the respiratory tract [28]; immunization of mice with antigen in the presence of cholera toxin gives rise to antigen-specific Tr1 cells [29, 30]; *Vitreoscilla filiformis* lysate can induce tolerogenic DCs and further drive the differentiation of murine Tr1 cells to suppress effector T cells and inflammation *in vivo* [31]; *Lactobacillus pentosus* KF340 can modulate DCs to promote Tr1 cell response, which can prevent systemic inflammation in a mouse model of atopic dermatitis [32]; oral administration of *Bifidobacterium breve* has been reported to mitigate intestinal inflammation in mice via the induction of Tr1 cells [33]; gut microbiota-derived short-chain fatty acids can enhance microbiota antigen-specific Tr1-like cell induction and inhibit murine colitis induced by dextran sulfate sodium [34].

In addition, pharmacological approaches, including PEGylated G-CSF [35], Bowman-Birk inhibitor, which is a soybean-derived serine protease inhibitor [36, 37], rapamycin combined with IL-10 [38] or G-CSF [39, 40], and a combination of immunosuppressive drugs vitamin D3 and dexamethasone [41], have been reported to induce Tr1 cells, therefore promote transplantation tolerance or suppress autoimmunity in mice *in vivo*.

IL-10 is known to be the primary cytokine that can drive the generation of both mouse and human Tr1 cells [24]. In addition, TGFβ secreted by CD4$^-$CD8$^-$CD11c$^+$ splenic DCs has been reported to induce the development of mouse Tr1 cells, which can mediate immune suppression *in vivo* [42]. In contrast, IFNα, but not TGFβ, can act synergistically with IL-10 to induce the generation of human Tr1 cells from naive CD4$^+$ T cells *in vitro* [52].

Of note, IL-10 administration alone failed to induce T cell tolerance in animal models of transplantation [68, 69] and autoimmune diseases [70, 71]. Results from these studies suggest that the induction of immune tolerance *in vivo* via Tr1 cell differentiation might require chronic antigen-specific stimulation in the presence of IL-10. Therefore, systemic administration of IL-10 alone may not be sufficient to control inflammatory immune response, and therefore fail to establish immune tolerance *in vivo* [24].

IL-10-producing macrophages with M2 phenotype have recently been reported to play an important role in immune tolerance via induction of Tr1 cells in the mouse model of allogeneic pancreatic islet transplantation [72]. In addition, multiple studies have suggested that IL-10 expressed by DCs plays a critical role in Tr1 cell induction. For example, IL-10-producing CD11c^{lo}CD45RB^{hi} plasmacytoid-like DCs in mouse lymph nodes and spleens have been reported to induce immune tolerance through the enhancement of Tr1 cell differentiation *in vivo* [73]. Repetitive stimulation of human naïve T cells with immature or mature DCs differentiated in the presence of exogenous IL-10 has also been reported to promote the differentiation of Tr1 cells *in vitro* [74, 75]. In addition, targeting self- and foreign antigens to myeloid DCs via C-type lectin receptor DC-ASGPR using antigen-antibody fusion proteins, which can stimulate IL-10 expression by DCs, has been shown to elicit antigen-specific immunosuppressive Tr1 cells from naïve and memory human CD4$^+$ T cells *in vitro* as well as in non-human primates *in vivo* [22, 55]. Furthermore, a subset of IL-10-producing human DCs, namely DC-10, has been discovered in the peripheral blood *in vivo* [75]. In addition, DC-10 can also be generated from monocytes *in vitro* in the presence of

IL-10 [75]. More importantly, DC-10 isolated from either peripheral blood or generated *in vitro* can induce antigen-specific Tr1 cells via an IL-10-dependent manner, which further supports the importance of DC-derived IL-10 in the induction of Tr1 cells [75–77].

In addition to DC-derived IL-10, inducible T cell costimulator ligand (ICOSL) expression by mouse pulmonary DCs plays an important role in the induction of IL-10-producing Tr1-like cells, as interruption of the ICOS–ICOSL signaling suppresses Tr1-like cell induction and blocks the development of tolerance to allergen in mouse *in vivo* [56]. Human plasmacytoid DCs (pDCs) express enhanced level of ICOSL upon maturation, which is known to promote the differentiation of IL-10-producing Tr1-like cells [57]. Furthermore, mature pDCs isolated from the peripheral blood of rheumatoid arthritis patients have high levels of indoleamine-2,3-dioxygenase (IDO), which can promote Tr1 cell differentiation [58]. However, whether ICOSL and IDO expression by pDCs is associated with IL-10 requires further investigation.

Of note, Tr1 cell development in the gut-associated lymphoid tissues in mice does not require IL-10 but mainly depends on TGFβ for their induction and/or maintenance, suggesting that other cytokine(s) could compensate for the absence of IL-10 to induce Tr1 cells [26]. Subsequent studies have further reported that IL-27 plays a critical role in inducing mouse Tr1 cells. Indeed, short-term activation of murine T cells in the presence of IL-27 results in the induction of Tr1 cells *in vitro* and *in vivo* [43–49]. In comparison, the role of IL-27 in inducing human Tr1 cells has been less studied. In humans, plasma IL-27 has been found to be correlated with peripheral $CD4^+IL\text{-}10^+$ T cells in Sjögren Syndrome patients [78]. Positive correlation between serum IL-27 and frequency of $CD4^+CD45RA^-CD49b^+LAG\text{-}3^+$ Tr1 cells in the peripheral blood has been reported in the severe forms of paracoccidioidomycosis [79]. In addition, IL-27 has been reported to induce IL-10-producing Tr1-like cells from human naïve $CD4^+$ T cells *in vitro* [50, 51], however, the suppressive function of these human Tr1-like cells requires further investigation.

In addition to cytokines mentioned above, retinoic acid (RA) has been reported to induce antigen-specific Tr1 cells in mouse *in vivo*, which is further enhanced by co-administration of IL-2 [61]. Immunization of mice with autoantigens in the presence of RA and IL-2 can suppress the development of autoimmunity in the mouse models of experimental autoimmune encephalomyelitis (EAE) and autoimmune uveitis [61]. In addition, RA by itself is sufficient to induce retinaldehyde dehydrogenase (RALDH) expression and endow human DCs with tolerogenic properties to elicit Tr1-like cells *in vitro* [62]. Aerobic glycolysis has also been reported to support Tr1 cell differentiation through a metabolic program controlled by hypoxia-inducible factor 1 alpha (HIF-1α) and aryl hydrocarbon receptor (AHR) [80], suggesting that both immunological and metabolic signals in a specific microenvironment can play pivotal roles in regulating Tr1 cell induction.

Furthermore, activation of human T cells via CD2 results in human Tr1 cell induction [63]. IL-10 is reported to downregulate the expression of costimulatory molecules CD80 and CD86 without affecting CD58/LFA-3 expression on antigen-presenting cells (APCs) [81]. Costimulation of human $CD4^+$ T cells via CD2 by its ligand CD58 induces the differentiation of Tr1 cells independently of IL-10 [63], suggesting that CD2 costimulation triggers an intrinsic signaling pathway resulting in Tr1 cell differentiation. In addition, costimulation of human naive $CD4^+$ cells through CD46 [64] or CD55 [65] can induce $CD4^+$ T cells to display a Tr1 cell phenotype. However, the precise mechanisms in which signals via CD46 and CD55 contribute to the induction of Tr1 cells remain to be determined.

4. Enigmatic lineage-defining transcription factor for Tr1 cells

So far, there is no master transcription regulator confirmed for either human or mouse Tr1 cells. Current understanding of mechanisms underlying the induction of Tr1 cells is mainly limited to IL-10 gene transactivation, and a number of transcription factors have been reported in this process (**Table 2**).

In mouse, IL-27 can promote IL-10 production by $CD4^+$ T cells through activation of STAT1 and STAT3 and drive the differentiation of Tr1 cells [44, 48, 82, 97]. Similarly, other cytokines that can activate STAT3, including IFNα and IL-6, have also been reported to promote Tr1 cells [52–54, 83].

In addition, IL-27-mediated signaling cascade through early growth response 2 (Egr-2) and B lymphocyte induced maturation protein-1 (Blimp-1) has been reported to play an important role in inducing mouse Tr1 cells. Retroviral gene transfer of Egr-2 can convert mouse naïve $CD4^+$ T cells into IL-10-producing and LAG-3-expressing antigen-specific immunosuppressive T cells *in vivo* [84]. Subsequent study has further revealed that IL-27 is sufficient to induce Egr-2, IL-10, and LAG-3 expression in mouse naïve $CD4^+$ T cells, whereas induction of IL-10 and Blimp-1 by

Markers	Mouse	References	Human	References
CD49b	+	[10]	+	[10]
LAG-3	+	[10]	+	[10]
CD226	+	[10]	+	[10]
Kinases and Transcription Factors				
STAT1	+	[48, 82]	+	[51]
STAT3	+	[44, 54]	+	[51, 83]
Egr-2	+	[47, 84]		
Blimp-1	+	[47, 85, 86]		
c-Maf	+	[45, 46, 86]	+	[60]
AHR	+	[45, 46]	+	[60]
IRF4	+	[87]	+	[60]
ITK	+	[87]	+	[87, 88]
Eomes	+	[89]	+	[89, 90]
ROR-α	+	[91]	+	[91]
IRF1	+	[92]		
BATF	+	[92]		
HIF-1α	+	[80]		
Mechanisms of Suppression				
Cytokines (IL-10 and TGFβ)	+	[24, 41]	+	[24, 74]
Killing of APCs (GzmB and perforin)			+	[93]
Cell-cell contact (CTLA-4 and PD-1)			+	[94, 95]
Metabolic disruption (CD39 and CD73)	+	[80]	+	[96]

Table 2.
Cellular and molecular features of mouse and human Tr1 Cells and their mechanisms of action.

IL-27 is dependent on Egr-2 [47]. Deficiency of Blimp-1 in mouse CD4$^+$ T cells results in impaired IL-10 production, whereas Blimp-1 overexpression has been reported to promote the Tr1 cell phenotype in effector T cells [98]. Blimp-1 is also found to promote IL-10 production by Tr1 cells in mouse models of malaria and visceral leishmaniasis [85]. A recent study has examined the transcriptional network driven by IL-27 across different mouse T cell subsets and identified multiple regulators of IL-10 expression [86]. Two central regulators, *Prdm1* (Blimp-1) and *Maf* (musculo-aponeurotic fibrosarcoma, c-Maf) are found to cooperatively drive the expression of signature genes induced by IL-27 in Tr1 cells and mediate IL-10 expression in all T helper cells [86]. More importantly, genetic depletion of *Prdm1* and *Maf* in T cells, but not either alone, results in the development of spontaneous colitis in mice, which underscores the importance of the crosstalk between *Prdm1* and *Maf* in the maintenance of immune homeostasis *in vivo* [86].

In addition, IL-27 is also reported to induce the expression of c-Maf, which acts in synergy with AHR, to promote IL-10 expression and differentiation of mouse Tr1 cells [45, 46]. Mice with impaired AHR signaling in CD4$^+$ T cells show lower IL-10 production and resistance to IL-27-mediated mitigation of EAE [46]. Furthermore, IL-27-driven c-Maf expression has been reported to transactivate IL-21 production [45, 46]. IL-21 by itself fails to induce Tr1 cells from native CD4$^+$ T cell, but it serves as an autocrine growth factor for the expansion as well as maintenance of Tr1 cells induced by IL-27, which is evidenced by the observation that loss of IL-21 signaling in CD4$^+$ T cells results in the inhibition of IL-27-driven generation of IL-10-producing T cells *in vitro* and in IL-21R-deficient mice *in vivo* [45]. Nonetheless, the detailed roles of IL-27 in the induction and activation of Tr1 cells in humans remain to be investigated.

Treatment of human naive T cells with activin-A, a member of the TGFβ superfamily, has been reported to induce the activation of interferon regulatory factor 4 (IRF4) [60]. IRF4, along with AHR and its binding partner, AHR nuclear translocator (ARNT), forms a tripartite transcription factor complex that is necessary for the differentiation and effector functions of human Tr1 cells [60]. In addition, IRF4 is also involved in the functional development of mouse Tr1 cells [87]. IL-2-inducible T-cell kinase (ITK) downstream of T cell receptor is found to serve a critical role for the activation of Ras/MAPK/IRF4 signal cascade, which further enables the functional development of Tr1 cells [87, 88]. Furthermore, adoptive transfer of human Tr1 cells induced by activin-A to a humanized mouse model of allergic asthma has been shown to provide the protection against major disease manifestations [60]. Activin-A is also reported to induce a population of antigen-specific IL-10-producing regulatory CD4$^+$ T cells, possibly representing Tr1 cells, which can protect against Th2-associated airway hyperresponsiveness and allergic airway disease in mice [59].

Other transcription factors, including Eomesodermin (Eomes) [89, 90], retinoic acid-related orphan receptor α (ROR-α) [91], have also been proposed to transactivate IL-10 expression in CD4$^+$ T cells and promote Tr1 cell differentiation. Furthermore, it has been reported that interferon regulatory factor 1 (IRF1) and basic leucine zipper ATF-like transcription factor (BATF) are induced early on during IL-27-induced Tr1 differentiation and act as pioneering factors for the differentiation of Tr1 cells in mouse [92]. BATF prepares the genomic landscape for the binding of additional transcription factors necessary for the development of Tr1 cells, and IRF1 specifically transactivates of the *Il10* gene for Tr1 cell differentiation in mice [92].

With the findings of these transcription factors in IL-10 gene transactivation, however, the lineage-defining transcription factor for mouse and human Tr1 cells is still elusive, which remains a key question to be answered in the study of Tr1 cells.

5. Phenotype of Tr1 cells

Though the immunosuppressive functions of Tr1 cells have been reported both *in vitro* and *in vivo*, their phenotype, in contrast, remains poorly defined. Coexpression of CD49b and LAG-3 has been proposed as the surface markers for both human and mouse Tr1 cells [10]. It has been known that CD49b is expressed on memory T cells [99], while LAG-3 is expressed on activated $CD4^+$ and $CD8^+$ T cells. In addition, high level of LAG-3 is also expressed by other immune cells, such as $Foxp3^+$ Tregs and IL-10-producing B cells [100, 101]. Therefore, the single use of CD49b or LAG-3 is not sufficient to define a pure population of functional Tr1 cells or separate these cells from other T helper cells or Treg cells. Of note, coexpression of CD49b and LAG-3 is found not limited to the $Foxp3^-$ Tr1 cells but is also observed in $Foxp3^+$ Tregs and $CD8^+$ T cells that produce IL-10 [102]. Indeed, IL-10-producing Tr1 cells, $Foxp3^+$ Tregs and $CD8^+$ T cells are all capable of co-expressing CD49b and LAG-3 *in vitro* when differentiated under IL-10-inducing conditions, and *in vivo* upon pathogenic encounter or infection in the pulmonary mucosa [102]. Therefore, it is highly recommended that a deliberate and precise gating strategy will need to be made to isolate $CD49b^+LAG\text{-}3^+$ memory Tr1 cells with the exclusion of B cells, $CD8^+$ and $Foxp3^+$ cells.

In addition to CD49b and LAG-3, Tr1 cells can express many other surface molecules, including PD-1, CTLA-4, TIGIT, TIM3, ICOS and CD226, as well as ectoenzymes CD39 and CD73, depending on the immune context [103, 104]. Although expression of such additional inhibitory receptors by Tr1 cells is generally in line with their immunosuppressive function, it is also necessary to realize that their expression is not specific to Tr1 cells.

6. Mechanism of Tr1 cells in immune suppression

The regulatory function of Tr1 cells requires their activation via TCR by cognate antigen recognition. In addition, Tr1 cells can also display bystander immunosuppressive activity to proximal T cells regardless of their antigen specificity. This indicates that activated Tr1 cells can regulate immune responses via both antigen-specific and non-specific manners (**Figure 1** and **Table 2**). Upon activation, human and mouse Tr1 cells secrete IL-10 and TGFβ [24, 41, 74], which suppress T cell responses directly and indirectly. IL-10 can limit the magnitude of immune responses by reducing the surface expression of MHC class II molecules [105–107], co-stimulatory molecules [108, 109], as well as the secretion of pro-inflammatory cytokines by APCs, followed by the suppression of effector T cell responses [104]. TGFβ expressed by Tr1 cells can also repress APC functions and inhibit T cell proliferation and cytokine production [52]. Granzyme B (GzmB) and perforin expressed by Tr1 cells can selectively kill APCs via both cognate and non-cognate mechanisms [93]. Cytolysis of the APCs can consequently suppress antigen-specific T cells and bystander T cells [110].

In addition to secretion of soluble factors including cytokines and enzymes, expression of inhibitory molecules PD-1 and CTLA-4 by Tr1 cells can repress effector T cells via cell contact-dependent manner, which is evidenced by that finding that blockade of CTLA-4 or PD-1 can decrease the suppressor activity of human Tr1 cells [94, 95]. Expression of ectoenzymes CD39 and CD73 [23, 103], though not exclusive to Tr1 cells as mentioned earlier, can facilitate Tr1-mediated suppression of effector T cells via metabolic disruption [80, 96]. In addition, IL-10-producing $Foxp3^-$ Tr1-like cells have also been reported to downregulate B cell antibody production due to low or no expression

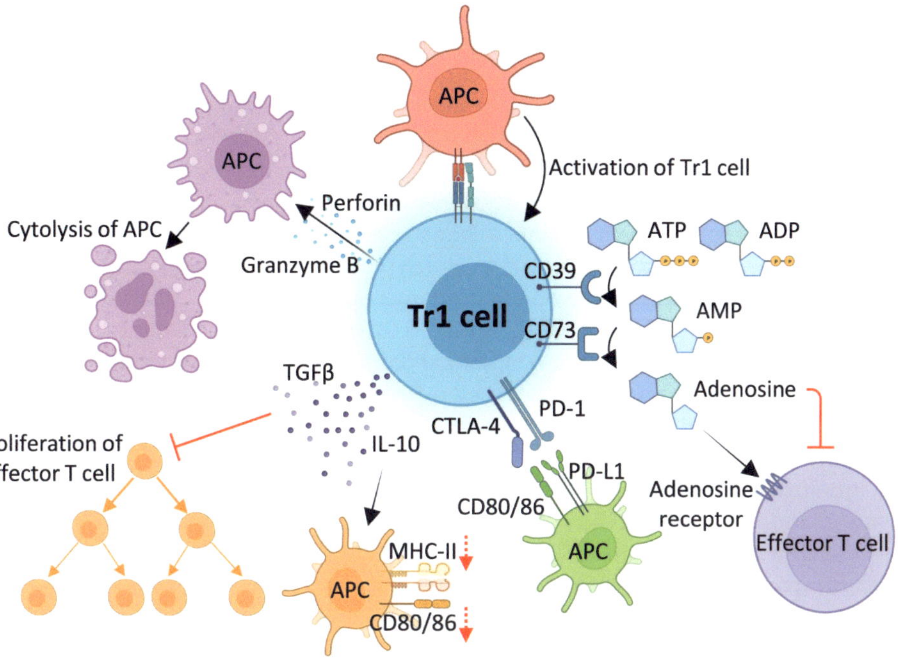

Figure 1.
Suppressive mechanisms of Tr1 cells. Upon activation, Tr1 cells suppress immune responses both directly and indirectly. The secretion of granzyme B and perforin can induce cytolysis of APCs, resulting in inhibition of both antigenspecific T cells and bystander T cells. TGFβ secreted by activated Tr1 cells can inhibit T cell proliferation and cytokine production while IL-10 can downregulate expression of MHC-II molecules, co-stimulatory molecules and production of pro-inflammatory cytokines by APCs. Tr1 cells can also inhibit APC-induced effector T cell activation via cell-cell contact involving CTLA-4 and PD-1. Ectoenzymes CD39 and CD73 expressed by Tr1 cells can mediate the suppression of effector T cells via metabolic disruption.

of CD40L [111]. Taken together, Tr1 cells can exhibit their immunosuppressive function through multiple mechanisms. In future studies, it would be of significance to investigate whether the suppressive mechanisms of Tr1 cells in different clinical settings are associated with the stage of disease progression and pathological microenvironment.

7. Therapeutic potentials of Tr1 cells

The immunoregulatory capacity of Tr1 cells has been tested in multiple different murine models of inflammatory diseases. $Foxp3^-$ Tr1 cells isolated from the intestine of $Il10^{eGFP}Foxp3^{RFP}$ double reporter mice have been shown to suppress colitis caused by the transfer of pathogenic Th17 cells in an IL-10-dependent manner *in vivo* [112]. In a mouse model of multiple sclerosis, transfer of OVA-specific Tr1 cells can prevent EAE development when antigen-specific Tr1 cells are activated by intracranial injection of OVA [41]. In a mouse GVHD model, Tr1 cells have been reported to constitute the most abundant regulatory population after allogeneic bone marrow transplantation [89]. More importantly, transfer of purified populations of Tr1 cells can significantly suppress GVHD and contribute significantly to transplant survival [89]. In acute and chronic collagen-induced arthritis mouse models, transfer of collagen type II-specific Tr1 cells can reduce the incidence and clinical symptoms of arthritis in both preventive and therapeutic settings, with a significant impact on collagen type

II-specific antibodies. Importantly, injection of collagen-specific Tr1 cells can significantly decrease the proliferation of antigen-specific effector T cells *in vivo* [113].

Human gliadin-specific Tr1 cell clones generated *in vitro* from the intestinal mucosa of celiac patients in remission have been shown to inhibit pathogenic T cell response to dietary gliadin [114]. With good manufacturing practice (GMP)-compatible protocols to differentiate and expand human Tr1 cells *in vitro*, Tr1 cells are also being used as a therapeutic product in clinical applications. IL-10-anergized donor T cells which contain Tr1 cells specific for recipient alloantigens generated *in vitro* have been tested in controlling GVHD in a clinical trial in which patients with high-risk/advanced stage hematologic malignancies received haplo-identical HSCT [115]. Patients had mild to moderate GVHD and showed rapid immune reconstitution [115]. Donor-derived T cells remained hyporesponsive to recipient alloantigens *in vitro* and an increase in cells with Tr1 cell signatures has been observed over time in recipients. Results from this study give the first indication of the feasibility of Tr1 cell-based immune therapy and show promise for the future use of Tr1 cells as treatment for hematologic malignancies and immune-related diseases [115]. T-allo10, as an improved cell product generated by stimulation of donor T cells with host-derived DC-10 in the presence of IL-10, contains a higher percentage of Tr1 cells (up to 15% of $CD49b^+$ $LAG\text{-}3^+$ Tr1 cells) [23], when compared with the previous generation of IL-10-anergized T cells (containing <5% Tr1 cells) used in the clinical trial mentioned above. The overall immunological outcome of T-allo10 in controlling GVHD in patients who received mismatched HSCT for the treatment of hematologic malignancies is still under investigation in an ongoing Phase I trial (ClinicalTrials.gov Identifier: NCT03198234).

In addition to GVHD in HSCT, the therapeutic effect of Tr1 cells in controlling graft rejection in solid organ transplantation is also being investigated. The protocol for generation of recipient-derived donor-specific Tr1 cells for kidney transplantation has been developed [116]. In addition, *Drosophila*-derived artificial APCs have been developed to induce antigen-specific Tr1 cells [66]. Schneider *Drosophila* cells transfected with a transmembrane form of a murine anti-human CD3 monoclonal antibody, together with human CD80 and CD58, as well as human IL-2 and IL-4, have been shown to expand a large number of antigen-specific Tr1 cells [66]. Using this method, Tr1 cells have also been tested to treat inflammatory disease in a phase I/IIa clinical study, in which OVA-specific Tr1 cell clones generated *in vitro* using artificial APCs have been adoptively transferred to patients with refractory Crohn's disease [67]. Patients were fed with OVA-enriched diet to ensure activation of OVA-specific Tr1 cells migrating to the gut. Administration of these OVA-specific Tr1 cells to patients with refractory Crohn's disease was well tolerated and had dose-related efficacy. Though the clinical effect was limited, reaching the maximum at 5 weeks after treatment and declining thereafter, the OVA-specific immune response correlated with clinical outcomes, supporting immunosuppressive function of OVA-specific Tr1 cells [67].

Induction of stable and sustained expression of IL-10 by conventional $CD4^+$ T cells has been developed as an alternative strategy to generate a large number of Tr1 cells. The lentiviral vectors encoding both human IL-10 gene and a marker GFP gene of selection have been tested to induce Tr1 cells [117]. It has been reported that lentiviral vector-mediated human IL-10 gene transfer converts conventional human $CD4^+$ T cells into Tr1-like cells, namely $CD4^{IL-10}$ cells. These cells resemble Tr1 cells phenotypically and functionally as they express large amount of IL-10, repress T cell responses *in vitro*, and more importantly, prevent xenogeneic GVHD development and progression *in vivo* [117]. Subsequent study has further reported that in addition to suppress T cell responses both *in vitro* and *in vivo*, $CD4^{IL-10}$ cells were also capable of killing myeloid leukemia cells in an HLA class I-dependent but Ag-independent manner [118]. This new generation of Tr1 cell product paves the way for adoptive cell therapy with Tr1 cells in patients undergoing allogeneic organ transplantation and HSCT [15, 119, 120].

Furthermore, utilization of artificial chimeric antigen receptors (CARs) to redirect regulatory T cell specificity towards pathogenic cell populations and antigens has also provided new insights in designing and implementing the next generation of Tr1 cells, CAR-Tr1 cells, for the treatment of transplantation rejection, autoimmunity, and leukemia [121–123]. In addition, genome editing techniques (including the application of CRISPR–Cas9) are under investigation to further enhance the specificity and immune regulatory functions of Tregs [122]. Together, all these progresses will certainly further increase the therapeutic value of Tr1 cells.

8. Conclusion

In the last two decades, the immune suppressive functions of Tr1 cells have been demonstrated both *in vitro* and *in vivo*. Studies from different groups have shown that Tr1 cells are able to prevent and constrain undesirable immune responses in different disease contexts, and therefore promote immune tolerance. These important discoveries have led to the idea that Tr1 cells could serve as a therapeutic product to promote and restore immune tolerance in transplantation, as well as in inflammatory and autoimmune diseases. The completed clinical trials have shown, to some extent, the safety of Tr1 cell-based therapy and further indicated the therapeutic potentials. Different methods are being developed to generate better Tr1 cell products. However, with all these advances, questions on Tr1 cells including whether they represent an established T cell lineage and whether Tr1 cells induced *in vitro* can maintain long-term immunoregulatory functions due to possible plasticity and repolarization *in vivo* remain unanswered. In addition, there is controversy over how well CD49b/LAG-3 surface co-expression signature defines circulating Tr1 cells in healthy individuals as many labs, including ours, struggle to use CD49b and LAG-3 to isolate Tr1 cells from bulk culture and it may not be the best approach when compared to the clinically applicable IL-10 cytokine capture method [14]. Future studies will need to elucidate the key molecules, including better and more stringent cell surface marker(s) as well as lineage-defining transcription factor(s) of Tr1 cells. In addition, a deeper and more comprehensive understanding on the biology of Tr1 cells is also necessary to deliver safer and more effective Tr1 cells that can be used to treat different diseases which require long-term regulation of inflammatory immune responses.

Acknowledgements

This study was supported by the Mayo Clinic Transplantation Research Center. We thank Dr. Matthew Wiest and Elizabeth Miller at Mayo Clinic for comments and suggestions on this review. **Figure 1** in this review was created with BioRender.com.

Conflict of interest

The authors declare no conflict of interest.

Author details

Chao Gu* and SangKon Oh
Department of Immunology, Mayo Clinic, Scottsdale, Arizona, United States

*Address all correspondence to: gu.chao@mayo.edu

References

[1] Chaplin DD. Overview of the immune response. The Journal of Allergy and Clinical Immunology. 2010;**125** (2 Suppl. 2):S3-S23

[2] Sakaguchi S, Mikami N, Wing JB, Tanaka A, Ichiyama K, Ohkura N. Regulatory T cells and human disease. Annual Review of Immunology. 2020;**38**:541-566

[3] Ligocki AJ, Niederkorn JY. Advances on non-CD4 + Foxp3+ T regulatory cells: CD8+, type 1, and double negative T regulatory cells in organ transplantation. Transplantation. 2015;**99**(8):1553-1559

[4] Rosser EC, Mauri C. Regulatory B cells: Origin, phenotype, and function. Immunity. 2015;**42**(4):607-612

[5] Mauri C, Menon M. Human regulatory B cells in health and disease: Therapeutic potential. The Journal of Clinical Investigation. 2017;**127**(3):772-779

[6] Deniz G, Erten G, Kucuksezer UC, Kocacik D, Karagiannidis C, Aktas E, et al. Regulatory NK cells suppress antigen-specific T cell responses. Journal of Immunology. 2008;**180**(2):850-857

[7] Neo SY, Yang Y, Record J, Ma R, Chen X, Chen Z, et al. CD73 immune checkpoint defines regulatory NK cells within the tumor microenvironment. The Journal of Clinical Investigation. 2020;**130**(3):1185-1198

[8] Gabrilovich DI, Nagaraj S. Myeloid-derived suppressor cells as regulators of the immune system. Nature Reviews Immunology. 2009;**9**(3):162-174

[9] Veglia F, Sanseviero E, Gabrilovich DI. Myeloid-derived suppressor cells in the era of increasing myeloid cell diversity. Nature Reviews Immunology. 2021;**21**(8):485-498

[10] Gagliani N, Magnani CF, Huber S, Gianolini ME, Pala M, Licona-Limon P, et al. Coexpression of CD49b and LAG-3 identifies human and mouse T regulatory type 1 cells. Nature Medicine. 2013;**19**(6):739-746

[11] Yao Y, Vent-Schmidt J, McGeough MD, Wong M, Hoffman HM, Steiner TS, et al. Tr1 cells, but not Foxp3(+) regulatory T cells, suppress NLRP3 inflammasome activation via an IL-10-dependent mechanism. Journal of Immunology. 2015;**195**(2):488-497

[12] Yu H, Gagliani N, Ishigame H, Huber S, Zhu S, Esplugues E, et al. Intestinal type 1 regulatory T cells migrate to periphery to suppress diabetogenic T cells and prevent diabetes development. Proceedings of the National Academy of Sciences of the United States of America. 2017;**114**(39):10443-10448

[13] Passerini L, Di Nunzio S, Gregori S, Gambineri E, Cecconi M, Seidel MG, et al. Functional type 1 regulatory T cells develop regardless of FOXP3 mutations in patients with IPEX syndrome. European Journal of Immunology. 2011;**41**(4):1120-1131

[14] Cook L, Stahl M, Han X, Nazli A, MacDonald KN, Wong MQ, et al. Suppressive and gut-reparative functions of human type 1 T regulatory cells. Gastroenterology. 2019;**157**(6):1584-1598

[15] Liu JMH, Chen P, Uyeda MJ, Cieniewicz B, Sayitoglu EC, Thomas BC, et al. Pre-clinical development and molecular characterization of an

engineered type 1 regulatory T-cell product suitable for immunotherapy. Cytotherapy. 2021;**23**(11):1017-1028

[16] Hori S, Nomura T, Sakaguchi S. Control of regulatory T cell development by the transcription factor Foxp3. Science. 2003;**299**(5609):1057-1061

[17] Fontenot JD, Gavin MA, Rudensky AY. Foxp3 programs the development and function of CD4+CD25+ regulatory T cells. Nature Immunology. 2003;**4**(4):330-336

[18] Hori S. FOXP3 as a master regulator of Treg cells. Nature Reviews. Immunology. 2021;**21**(10):618-619

[19] Allan SE, Alstad AN, Merindol N, Crellin NK, Amendola M, Bacchetta R, et al. Generation of potent and stable human CD4(+) T regulatory cells by activation-independent expression of FOXP3. Molecular Therapy. 2008;**16**(1):194-202

[20] Lu L, Barbi J, Pan F. The regulation of immune tolerance by FOXP3. Nature Reviews Immunology. 2017;**17**(11):703-717

[21] O'Garra A, Vieira PL, Vieira P, Goldfeld AE. IL-10-producing and naturally occurring CD4(+) Tregs: Limiting collateral damage. Journal of Clinical Investigation. 2004;**114**(10):1372-1378

[22] Li D, Romain G, Flamar AL, Duluc D, Dullaers M, Li XH, et al. Targeting self- and foreign antigens to dendritic cells via DC-ASGPR generates IL-10-producing suppressive CD4+ T cells. The Journal of Experimental Medicine. 2012;**209**(1):109-121

[23] Roncarolo MG, Gregori S, Bacchetta R, Battaglia M, Gagliani N. The biology of T regulatory type 1 cells and their therapeutic application in immune-mediated diseases. Immunity. 2018;**49**(6):1004-1019

[24] Groux H, O'Garra A, Bigler M, Rouleau M, Antonenko S, de Vries JE, et al. A CD4+ T-cell subset inhibits antigen-specific T-cell responses and prevents colitis. Nature. 1997;**389**(6652):737-742

[25] Kamanaka M, Kim ST, Wan YY, Sutterwala FS, Lara-Tejero M, Galan JE, et al. Expression of interleukin-10 in intestinal lymphocytes detected by an interleukin-10 reporter knockin tiger mouse. Immunity. 2006;**25**(6):941-952

[26] Maynard CL, Harrington LE, Janowski KM, Oliver JR, Zindl CL, Rudensky AY, et al. Regulatory T cells expressing interleukin 10 develop from Foxp3+ and Foxp3- precursor cells in the absence of interleukin 10. Nature Immunology. 2007;**8**(9):931-941

[27] Zuany-Amorim C, Sawicka E, Manlius C, Le Moine A, Brunet LR, Kemeny DM, et al. Suppression of airway eosinophilia by killed Mycobacterium vaccae-induced allergen-specific regulatory T-cells. Nature Medicine. 2002;**8**(6):625-629

[28] McGuirk P, McCann C, Mills KH. Pathogen-specific T regulatory 1 cells induced in the respiratory tract by a bacterial molecule that stimulates interleukin 10 production by dendritic cells: A novel strategy for evasion of protective T helper type 1 responses by Bordetella pertussis. The Journal of Experimental Medicine. 2002;**195**(2):221-231

[29] Lavelle EC, McNeela E, Armstrong ME, Leavy O, Higgins SC, Mills KH. Cholera toxin promotes the induction of regulatory T cells specific for bystander antigens by modulating

dendritic cell activation. Journal of Immunology. 2003;**171**(5):2384-2392

[30] Lavelle EC, Jarnicki A, McNeela E, Armstrong ME, Higgins SC, Leavy O, et al. Effects of cholera toxin on innate and adaptive immunity and its application as an immunomodulatory agent. Journal of Leukocyte Biology. 2004;**75**(5):756-763

[31] Volz T, Skabytska Y, Guenova E, Chen KM, Frick JS, Kirschning CJ, et al. Nonpathogenic bacteria alleviating atopic dermatitis inflammation induce IL-10-producing dendritic cells and regulatory Tr1 cells. Journal of Investigative Dermatology. 2014;**134**(1):96-104

[32] Kim JE, Sharma A, Sharma G, Lee SY, Shin HS, Rudra D, et al. Lactobacillus pentosus modulates immune response by inducing IL-10 producing Tr1 cells. Immune Network. 2019;**19**(6)

[33] Jeon SG, Kayama H, Ueda Y, Takahashi T, Asahara T, Tsuji H, et al. Probiotic bifidobacterium breve induces IL-10-producing Tr1 cells in the colon. PLoS Pathogens. 2012;**8**(5)

[34] Sun MM, Wu W, Chen L, Yang WJ, Huang XS, Ma CY, et al. Microbiota-derived short-chain fatty acids promote Th1 cell IL-10 production to maintain intestinal homeostasis. Nature Communications. 2018;**9**

[35] Morris ES, MacDonald KP, Rowe V, Johnson DH, Banovic T, Clouston AD, et al. Donor treatment with pegylated G-CSF augments the generation of IL-10-producing regulatory T cells and promotes transplantation tolerance. Blood. 2004;**103**(9):3573-3581

[36] Dai H, Ciric B, Zhang GX, Rostami A. Interleukin-10 plays a crucial role in suppression of experimental autoimmune encephalomyelitis by Bowman-Birk inhibitor. Journal of Neuroimmunology. 2012;**245**(1-2):1-7

[37] Safavi F, Thome R, Li ZC, Wang LM, Rasouli J, Ciric B, et al. A serine protease inhibitor induces type 1 regulatory T cells through IFN-gamma/STAT1 signaling. Cellular & Molecular Immunology. 2020;**17**(9):1004-1006

[38] Battaglia M, Stabilini A, Draghici E, Gregori S, Mocchetti C, Bonifacio E, et al. Rapamycin and interleukin-10 treatment induces T regulatory type 1 cells that mediate antigen-specific transplantation tolerance. Diabetes. 2006;**55**(1):40-49

[39] Gagliani N, Gregori S, Jofra T, Valle A, Stabilini A, Rothstein DM, et al. Rapamycin combined with anti-CD45RB mAb and IL-10 or with G-CSF induces tolerance in a stringent mouse model of islet transplantation. PLoS One. 2011;**6**(12)

[40] Fousteri G, Jofra T, Di Fonte R, Battaglia M. Combination of an antigen-specific therapy and an immunomodulatory treatment to simultaneous block recurrent autoimmunity and alloreactivity in non-obese diabetic mice. PLoS One. 2015;**10**(6)

[41] Barrat FJ, Cua DJ, Boonstra A, Richards DF, Crain C, Savelkoul HF, et al. In vitro generation of interleukin 10-producing regulatory CD4(+) T cells is induced by immunosuppressive drugs and inhibited by T helper type 1 (Th1)- and Th2-inducing cytokines. Journal of Experimental Medicine. 2002;**195**(5):603-616

[42] Zhang XS, Huang H, Yuan JY, Sun DM, Hou WS, Gordon J, et al. CD4(−)8(−) dendritic cells prime CD4(+) T regulatory 1 cells to suppress antitumor immunity. Journal of Immunology. 2005;**175**(5):2931-2937

[43] Awasthi A, Carrier Y, Peron JP, Bettelli E, Kamanaka M, Flavell RA, et al. A dominant function for interleukin 27 in generating interleukin 10-producing anti-inflammatory T cells. Nature Immunology. 2007;**8**(12):1380-1389

[44] Stumhofer JS, Silver JS, Laurence A, Porrett PM, Harris TH, Turka LA, et al. Interleukins 27 and 6 induce STAT3-mediated T cell production of interleukin 10. Nature Immunology. 2007;**8**(12):1363-1371

[45] Pot C, Jin H, Awasthi A, Liu SM, Lai CY, Madan R, et al. Cutting edge: IL-27 induces the transcription factor c-Maf, cytokine IL-21, and the costimulatory receptor ICOS that coordinately act together to promote differentiation of IL-10-producing Tr1 cells. Journal of Immunology. 2009;**183**(2):797-801

[46] Apetoh L, Quintana FJ, Pot C, Joller N, Xiao S, Kumar D, et al. The aryl hydrocarbon receptor interacts with c-Maf to promote the differentiation of type 1 regulatory T cells induced by IL-27. Nature Immunology. 2010;**11**(9):854-U112

[47] Iwasaki Y, Fujio K, Okamura T, Yanai A, Sumitomo S, Shoda H, et al. Egr-2 transcription factor is required for Blimp-1-mediated IL-10 production in IL-27-stimulated CD4+ T cells. European Journal of Immunology. 2013;**43**(4):1063-1073

[48] Wu C, Pot C, Apetoh L, Thalhamer T, Zhu B, Murugaiyan G, et al. Metallothioneins negatively regulate IL-27-induced type 1 regulatory T-cell differentiation. Proceedings of the National Academy of Sciences of the United States of America. 2013;**110**(19):7802-7807

[49] Nadya NA, Tezuka H, Ohteki T, Matsuda S, Azuma M, Nagai S. PI3K-Akt pathway enhances the differentiation of interleukin-27-induced type 1 regulatory T cells. Immunology. 2017;**152**(3):507-516

[50] Murugaiyan G, Mittal A, Lopez-Diego R, Maier LM, Anderson DE, Weiner HL. IL-27 is a key regulator of IL-10 and IL-17 production by human CD4(+) T cells. Journal of Immunology. 2009;**183**(4):2435-2443

[51] Wang H, Meng R, Li ZT, Yang BY, Liu Y, Huang FY, et al. IL-27 induces the differentiation of Tr1-like cells from human naive CD4(+) T cells via the phosphorylation of STAT1 and STAT3. Immunology Letters. 2011;**136**(1):21-28

[52] Levings MK, Sangregorio R, Galbiati F, Squadrone S, Malefyt RD, Roncarolo MG. IFN-alpha and IL-10 induce the differentiation of human type 1 T regulatory cells. Journal of Immunology. 2001;**166**(9):5530-5539

[53] Corre B, Perrier J, El Khouri M, Cerboni S, Pellegrini S, Michel F. Type I interferon potentiates T-cell receptor mediated induction of IL-10-producing CD4(+) T cells. European Journal of Immunology. 2013;**43**(10):2730-2740

[54] Jin JO, Han XZ, Yu Q. Interleukin-6 induces the generation of IL-10-producing Tr1 cells and suppresses autoimmune tissue inflammation. Journal of Autoimmunity. 2013;**40**:28-44

[55] Gu C, Wang L, Zurawski S, Oh S. Signaling cascade through DC-ASGPR induces transcriptionally active CREB for IL-10 induction and immune regulation. Journal of Immunology. 2019;**203**(2):389-399

[56] Akbari O, Freeman GJ, Meyer EH, Greenfield EA, Chang TT, Sharpe AH, et al. Antigen-specific regulatory T cells develop via the ICOS-ICOS-ligand pathway and inhibit allergen-induced

airway hyperreactivity. Nature Medicine. 2002;**8**(9):1024-1032

[57] Ito T, Yang M, Wang YH, Lande R, Gregorio J, Perng OA, et al. Plasmacytoid dendritic cells prime IL-10-producing T regulatory cells by inducible costimulator ligand. Journal of Experimental Medicine. 2007;**204**(1):105-115

[58] Kavousanaki M, Makrigiannakis A, Boumpas D, Verginis P. Novel role of plasmacytoid dendritic cells in humans induction of interleukin-10-producing treg cells by plasmacytoid dendritic cells in patients with rheumatoid arthritis responding to therapy. Arthritis Rheum-Us. 2010;**62**(1):53-63

[59] Semitekolou M, Alissafi T, Aggelakopoulou M, Kourepini E, Kariyawasam HH, Kay AB, et al. Activin-A induces regulatory T cells that suppress T helper cell immune responses and protect from allergic airway disease. Journal of Experimental Medicine. 2009;**206**(8):1769-1785

[60] Tousa S, Semitekolou M, Morianos I, Banos A, Trochoutsou AI, Brodie TM, et al. Activin-A co-opts IRF4 and AhR signaling to induce human regulatory T cells that restrain asthmatic responses. Proceedings of the National Academy of Sciences of the United States of America. 2017;**114**(14):E2891-EE900

[61] Raverdeau M, Christofi M, Malara A, Wilk MM, Misiak A, Kuffova L, et al. Retinoic acid-induced autoantigen-specific type 1 regulatory T cells suppress autoimmunity. EMBO Reports. 2019;**20**(5)

[62] Bakdash G, Vogelpoel LTC, van Capel TMM, Kapsenberg ML, de Jong EC. Retinoic acid primes human dendritic cells to induce gut-homing, IL-10-producing regulatory T cells. Mucosal Immunology. 2015;**8**(2):265-278

[63] Wakkach A, Cottrez F, Groux H. Differentiation of regulatory T cells 1 is induced by CD2 costimulation. Journal of Immunology. 2001;**167**(6):3107-3113

[64] Kemper C, Chan AC, Green JM, Brett KA, Murphy KM, Atkinson JP. Activation of human CD4(+) cells with CD3 and CD46 induces a T-regulatory cell 1 phenotype. Nature. 2003;**421**(6921): 388-392

[65] Sutavani RV, Bradley RG, Ramage JM, Jackson AM, Durrant LG, Spendlove I. CD55 Costimulation induces differentiation of a discrete T regulatory type 1 cell population with a stable phenotype. Journal of Immunology. 2013;**191**(12):5895-5903

[66] Brun V, Bastian H, Neveu V, Foussat A. Clinical grade production of IL-10 producing regulatory Tr1 lymphocytes for cell therapy of chronic inflammatory diseases. International Immunopharmacology. 2009;**9**(5):609-613

[67] Desreumaux P, Foussat A, Allez M, Beaugerie L, Hebuterne X, Bouhnik Y, et al. Safety and efficacy of antigen-specific regulatory T-cell therapy for patients with refractory Crohn's disease. Gastroenterology. 2012;**143**(5):1207-17 e2

[68] Lee MS, Wogensen L, Shizuru J, Oldstone MB, Sarvetnick N. Pancreatic islet production of murine interleukin-10 does not inhibit immune-mediated tissue destruction. The Journal of Clinical Investigation. 1994;**93**(3):1332-1338

[69] Blazar BR, Taylor PA, Smith S, Vallera DA. Interleukin-10 administration decreases survival in murine recipients of major histocompatibility complex disparate donor bone marrow grafts. Blood. 1995;**85**(3):842-851

[70] Nisitani S, Tsubata T, Murakami M, Honjo T. Administration of interleukin-5 or −10 activates peritoneal B-1 cells and induces autoimmune hemolytic anemia in anti-erythrocyte autoantibody-transgenic mice. European Journal of Immunology. 1995;**25**(11):3047-3052

[71] Cannella B, Gao YL, Brosnan C, Raine CS. IL-10 fails to abrogate experimental autoimmune encephalomyelitis. Journal of Neuroscience Research. 1996;**45**(6):735-746

[72] Mfarrej B, Jofra T, Morsiani C, Gagliani N, Fousteri G, Battaglia M. Key role of macrophages in tolerance induction via T regulatory type 1 (Tr1) cells. Clinical and Experimental Immunology. 2020;**201**(2):222-230

[73] Wakkach A, Fournier N, Brun V, Breittmayer JP, Cottrez F, Groux H. Characterization of dendritic cells that induce tolerance and T regulatory 1 cell differentiation in vivo. Immunity. 2003;**18**(5):605-617

[74] Levings MK, Gregori S, Tresoldi E, Cazzaniga S, Bonini C, Roncarolo MG. Differentiation of Tr1 cells by immature dendritic cells requires IL-10 but not CD25+CD4+ Tr cells. Blood. 2005;**105**(3):1162-1169

[75] Gregori S, Tomasoni D, Pacciani V, Scirpoli M, Battaglia M, Magnani CF, et al. Differentiation of type 1 T regulatory cells (Tr1) by tolerogenic DC-10 requires the IL-10-dependent ILT4/HLA-G pathway. Blood. 2010;**116**(6):935-944

[76] Amodio G, Gregori S. Human tolerogenic DC-10: Perspectives for clinical applications. Transplantation Research. 2012;**1**(1):14

[77] Amodio G, Comi M, Tomasoni D, Gianolini ME, Rizzo R, LeMaoult J, et al. HLA-G expression levels influence the tolerogenic activity of human DC-10. Haematologica. 2015;**100**(4):548-557

[78] Qi JJ, Zhang ZY, Tang XJ, Li WC, Chen WW, Yao GH. IL-27 regulated CD4(+)IL-10(+)T cells in experimental sjogren syndrome. Frontiers in Immunology. 2020;**11**

[79] Genaro LM, Coser LD, Justo AD, de Castro LF, Barreto AKF, Rizzato AE, et al. Association between IL-27 and Tr1 cells in severe form of paracoccidioidomycosis. Cytokine. 2020;**127**

[80] Mascanfroni ID, Takenaka MC, Yeste A, Patel B, Wu Y, Kenison JE, et al. Metabolic control of type 1 regulatory T cell differentiation by AHR and HIF1-alpha. Nature Medicine. 2015;**21**(6):638-646

[81] Groux H, Bigler M, de Vries JE, Roncarolo MG. Inhibitory and stimulatory effects of IL-10 on human CD8(+) T cells. Journal of Immunology. 1998;**160**(7):3188-3193

[82] Fitzgerald DC, Zhang GX, El-Behi M, Fonseca-Kelly Z, Li H, Yu S, et al. Suppression of autoimmune inflammation of the central nervous system by interleukin 10 secreted by interleukin 27-stimulated T cells. Nature Immunology. 2007;**8**(12):1372-1379

[83] Ahlers J, Mantei A, Lozza L, Staber M, Heinrich F, Bacher P, et al. A Notch/STAT3-driven Blimp-1/c-Maf-dependent molecular switch induces IL-10 expression in human CD4(+) T cells and is defective in Crohn's disease patients. Mucosal Immunology. 2022;**15**(3):480-490

[84] Okamura T, Fujio K, Shibuya M, Sumitomo S, Shoda H, Sakaguchi S, et al. CD4(+)CD25(−)LAG3(+) regulatory

T cells controlled by the transcription factor Egr-2. Proceedings of the National Academy of Sciences of the United States of America. 2009;**106**(33):13974-13979

[85] de Oca MM, Kumar R, Rivera FD, Amante FH, Sheel M, Faleiro RJ, et al. Blimp-1-dependent IL-10 production by Tr1 cells regulates TNF-mediated tissue pathology. PLoS Pathogens. 2016;**12**(1)

[86] Zhang H, Madi A, Yosef N, Chihara N, Awasthi A, Pot C, et al. An IL-27-driven transcriptional network identifies regulators of IL-10 expression across T helper cell subsets. Cell Reports. 2020;**33**(8):108433

[87] Huang WS, Solouki S, Koylass N, Zheng SG, August A. ITK signalling via the Ras/IRF4 pathway regulates the development and function of Tr1 cells. Nature Communications. 2017;**8**

[88] McGee MC, August A, Huang WS. TCR/ITK signaling in type 1 regulatory T cells. Advances in Experimental Medicine and Biology. 2021;**1278**:115-124

[89] Zhang P, Lee JS, Gartlan KH, Schuster IS, Comerford I, Varelias A, et al. Eomesodermin promotes the development of type 1 regulatory T (TR1) cells. Science Immunology. 2017;**2**(10)

[90] Gruarin P, Maglie S, De Simone M, Haringer B, Vasco C, Ranzani V, et al. Eomesodermin controls a unique differentiation program in human IL-10 and IFN-gamma coproducing regulatory T cells. European Journal of Immunology. 2019;**49**(1):96-111

[91] Farez MF, Mascanfroni ID, Mendez-Huergo SP, Yeste A, Murugaiyan G, Garo LP, et al. Melatonin contributes to the seasonality of multiple sclerosis relapses. Cell. 2015;**162**(6):1338-1352

[92] Karwacz K, Miraldi ER, Pokrovskii M, Madi A, Yosef N, Wortman I, et al. Critical role of IRF1 and BATF in forming chromatin landscape during type 1 regulatory cell differentiation. Nature Immunology. 2017;**18**(4):412

[93] Magnani CF, Alberigo G, Bacchetta R, Serafini G, Andreani M, Roncarolo MG, et al. Killing of myeloid APCs via HLA class I, CD2 and CD226 defines a novel mechanism of suppression by human Tr1 cells. European Journal of Immunology. 2011;**41**(6):1652-1662

[94] Akdis M, Verhagen J, Taylor A, Karamloo F, Karagiannidis C, Crameri R, et al. Immune responses in healthy and allergic individuals are characterized by a fine balance between allergen-specific T regulatory 1 and T helper 2 cells. Journal of Experimental Medicine. 2004;**199**(11):1567-1575

[95] Chen PP, Cepika AM, Agarwal-Hashmi R, Saini G, Uyeda MJ, Louis DM, et al. Alloantigen-specific type 1 regulatory T cells suppress through CTLA-4 and PD-1 pathways and persist long-term in patients. Science Translational Medicine. 2021;**13**(617)

[96] Mandapathil M, Szczepanski MJ, Szajnik M, Ren J, Jackson EK, Johnson JT, et al. Adenosine and prostaglandin E-2 cooperate in the suppression of immune responses mediated by adaptive regulatory T cells. The Journal of Biological Chemistry. 2010;**285**(36):27571-27580

[97] Hirahara K, Onodera A, Villarino AV, Bonelli M, Sciume G, Laurence A, et al. Asymmetric action of STAT transcription factors drives transcriptional outputs and cytokine specificity. Immunity. 2015;**42**(5):877-889

[98] Heinemann C, Heink S, Petermann F, Vasanthakumar A, Rothhammer V, Doorduijn E, et al. IL-27 and IL-12 oppose pro-inflammatory IL-23 in CD4+ T cells by inducing Blimp1. Nature Communications. 2014;**5**:3770

[99] Kassiotis G, Gray D, Kiafard Z, Zwirner J, Stockinger B. Functional specialization of memory Th cells revealed by expression of integrin CD49b. Journal of Immunology. 2006;**177**(2):968-975

[100] Lino AC, Dang VD, Lampropoulou V, Welle A, Joedicke J, Pohar J, et al. LAG-3 inhibitory receptor expression identifies immunosuppressive natural regulatory plasma cells. Immunity. 2018;**49**(1):120

[101] Camisaschi C, Casati C, Rini F, Perego M, De Filippo A, Rini F, et al. LAG-3 expression defines a subset of CD4(+)CD25(high)Foxp3(+) regulatory T cells that are expanded at tumor sites. Journal of Immunology. 2010;**184**(11):6545-6551

[102] Huang W, Solouki S, Carter C, Zheng SG, August A. Beyond type 1 regulatory T cells: Co-expression of LAG3 and CD49b in IL-10-producing T cell lineages. Frontiers in Immunology. 2018;**9**:2625

[103] Anderson AC, Joller N, Kuchroo VK. Lag-3, Tim-3, and TIGIT: Co-inhibitory receptors with specialized functions in immune regulation. Immunity. 2016;**44**(5):989-1004

[104] Roncarolo MG, Gregori S, Bacchetta R, Battaglia M. Tr1 cells and the counter-regulation of immunity: Natural mechanisms and therapeutic applications. Current Topics in Microbiology and Immunology. 2014;**380**:39-68

[105] Malefyt RD, Haanen J, Spits H, Roncarolo MG, Tevelde A, Figdor C, et al. Interleukin-10 (Il-10) and viral-Il-10 strongly reduce antigen-specific human T-cell proliferation by diminishing the antigen-presenting capacity of monocytes via down-regulation of class-Ii major histocompatibility complex expression. Journal of Experimental Medicine. 1991;**174**(4):915-924

[106] Koppelman B, Neefjes JJ, de Vries JE, Malefyt RD. Interleukin-10 down-regulates MHC class II alpha beta peptide complexes at the plasma membrane of monocytes by affecting arrival and recycling. Immunity. 1997;**7**(6):861-871

[107] Thibodeau J, Bourgeois-Daigneault MC, Huppe G, Tremblay J, Aumont A, Houde M, et al. Interleukin-10-induced MARCH1 mediates intracellular sequestration of MHC class II in monocytes. European Journal of Immunology. 2008;**38**(5):1225-1230

[108] Ding L, Linsley PS, Huang LY, Germain RN, Shevach EM. Il-10 inhibits macrophage costimulatory activity by selectively inhibiting the up-regulation of B7 expression. Journal of Immunology. 1993;**151**(3):1224-1234

[109] Chang CH, Furue M, Tamaki K. B7-1 expression of langerhans cells is up-regulated by proinflammatory cytokines, and is down-regulated by interferon-gamma or by interleukin-10. European Journal of Immunology. 1995;**25**(2):394-398

[110] Tree TI, Lawson J, Edwards H, Skowera A, Arif S, Roep BO, et al. Naturally arising human CD4 T-cells that recognize islet autoantigens and secrete interleukin-10 regulate proinflammatory T-cell responses via linked suppression. Diabetes. 2010;**59**(6):1451-1460

[111] Facciotti F, Gagliani N, Haringer B, Alfen JS, Penatti A, Maglie S, et al.

IL-10-producing forkhead box protein 3-negative regulatory T cells inhibit B-cell responses and are involved in systemic lupus erythematosus. The Journal of Allergy and Clinical Immunology. 2016;**137**(1):318-21 e5

[112] Huber S, Gagliani N, Esplugues E, O'Connor W, Huber FJ, Chaudhry A, et al. Th17 cells express interleukin-10 receptor and are controlled by Foxp3(−) and Foxp3(+) regulatory CD4(+) T cells in an interleukin-10-dependent manner. Immunity. 2011;**34**(4):554-565

[113] Asnagli H, Martire D, Belmonte N, Quentin J, Bastian H, Boucard-Jourdin M, et al. Type 1 regulatory T cells specific for collagen type II as an efficient cell-based therapy in arthritis. Arthritis Research & Therapy. 2014;**16**(3)

[114] Gianfrani C, Levings MK, Sartirana C, Mazzarella G, Barba G, Zanzi D, et al. Gliadin-specific type 1 regulatory T cells from the intestinal mucosa of treated celiac patients inhibit pathogenic T cells. Journal of Immunology. 2006;**177**(6):4178-4186

[115] Bacchetta R, Lucarelli B, Sartirana C, Gregori S, Stanghellini MTL, Miqueu P, et al. Immunological outcome in haploicentical-HSC transplanted patients treated with IL-10-anergizec donor T cells. Frontiers in Immunology. 2014;**5**

[116] Mfarrej B, Tresoldi E, Stabilini A, Paganelli A, Caldara R, Secchi A, et al. Generation of donor-specific Tr1 cells to be used after kidney transplantation and definition of the timing of their in vivo infusion in the presence of immunosuppression. Journal of Translational Medicine. 2017;**15**

[117] Andolfi G, Fousteri G, Rossetti M, Magnani CF, Jofra T, Locafaro G, et al. Enforced IL-10 expression confers type 1 regulatory T cell (Tr1) phenotype and function to human CD4(+) T cells. Molecular Therapy. 2012;**20**(9):1778-1790

[118] Locafaro G, Andolfi G, Russo F, Cesana L, Spinelli A, Camisa B, et al. IL-10-Engineered human CD4(+) Tr1 cells eliminate myeloid leukemia in an HLA class I-dependent mechanism. Molecular Therapy. 2017;**25**(10):2254-2269

[119] Gregori S, Roncarolo MG. Engineered T regulatory type 1 cells for clinical application. Frontiers in Immunology. 2018;**9**

[120] Cieniewicz B, Uyeda MJ, Chen PG, Sayitoglu EC, Liu JMH, Andolfi G, et al. Engineered type 1 regulatory T cells designed for clinical use kill primary pediatric acute myeloid leukemia cells. Haematologica. 2021;**106**(10):2588-2597

[121] Dawson NAJ, Levings MK. Antigen-specific regulatory T cells: Are police CARs the answer? Translational Research. 2017;**187**:53-58

[122] Ferreira LMR, Muller YD, Bluestone JA, Tang QZ. Next-generation regulatory T cell therapy. Nature Reviews. Drug Discovery. 2019;**18**(10):749-769

[123] Harris DT, Kranz DM. Adoptive T cell therapies: A comparison of T cell receptors and chimeric antigen receptors. Trends in Pharmacological Sciences. 2016;**37**(3):220-230

Chapter 2

Function and Therapeutic Intervention of Regulatory T Cells in Immune Regulation

Yang Liu, Tiezheng Hou and Huiqin Hao

Abstract

Although the detailed mechanisms of regulatory T cells (Tregs) in regulating immune responses have not been completely clarified yet, Tregs therapy on autoimmune diseases and organ transplantation is making robust progress, along with the gradually enhancing knowledge of the Tregs function. In this chapter, on the basis of summarizing the immunomodulatory functions of Tregs, we reviewed the latest scientific progress and status of our understanding, as well as the prospect of stimulation and expansion of Tregs *in vivo* and *in vitro* followed by adoptive transfer or autologous cell therapy in animal models and clinical trials, respectively. Moreover, we also assessed the current technological limitation and potential side effects of polyclonal and antigen-specific Tregs-based approaches and techniques, to promote the development of rescue, revive, or rejuvenate Tregs in the therapeutic intervention to treat autoimmune diseases and transplantation.

Keywords: regulatory T cell, polyclonal Treg, antigen-specific Treg, therapeutic intervention, transplantation, autoimmune disease

1. Introduction

Regulatory T cells (Tregs), as a subgroup of T cells with immunosuppressive function, were first reported in 1970s by Gershon and Kondo [1]. According to the developmental origin, Tregs can be broadly classified into two groups. Tregs that grow in the thymus are called natural (nTregs) or thymic (tTregs) Tregs, and that develop at the periphery by specific stimuli of conventional $CD4^+$ T cells are termed peripheral Treg (pTregs). When Tregs are induced by specific factors, such as interleukin-2 (IL-2) and transforming growth factor (TGF)-β, *in vitro* are called induced Tregs (iTregs) [2]. At present, Tregs have emerged as a vital part in understanding the immune response to pathogens, controlling the development of allergies, transplantation, and autoimmune diseases, as well as in the application of treating tumors, since their "re-discovery" more than 20 years before [3]. However, the detailed mechanisms of Tregs in regulating both innate and adaptive immune responses are still not completely understood. In this chapter, we will review the latest scientific progress and status on our understanding and prospect of stimulation and expansion of Tregs *in vivo* and *in vitro* followed by adoptive transfer or autologous cell therapy in animal models and clinical trials.

We will also assess the current technological limitation and potential side effects of polyclonal and antigen-specific Tregs-cell-based approaches and techniques.

2. Tregs function

2.1 Loss of Tregs and development of autoimmune diseases

Every manifestation stemming from Tregs paucity highlights a vital function of Tregs in preventing fatal autoimmune inflammation. The immunosuppressive function of Tregs is mainly dependent on continuous expression of the transcription factor forkhead box protein 3 (Foxp3), which is a critical regulator of $CD4^{+}CD25^{+}$ Tregs development and function. Loss function of Foxp3 results in a fatal autoimmune disease featuring all known types of inflammatory responses. Studies have demonstrated that the typical or fatal autoimmune responses that occurred in the Foxp3-mutant scurfy mice or Foxp3-null mice are related to the deficiency of $CD4^{+}CD25^{+}$ Tregs, but not to the cell-intrinsic dysfunction of $CD4^{+}CD25^{-}$ T cells. When being transferred into the neonatal Foxp3-deficient mice, Tregs can preferentially expand and control the development of autoimmune disease. Furthermore, ectopic expression of Foxp3 can confer suppressor function on peripheral $CD4^{+}CD25^{-}$ T cells [4]. Even in severely diseased mice, by reinstating Foxp3 protein expression and suppressor function in cells expressing a reversible Foxp3 null allele, the rescued Tregs normalized immune activation, quelled severe tissue inflammation, reversed fatal autoimmune disease, and provided long-term protection against them. It is indicated that Tregs are capable of resetting the immune homeostasis in broad-spectrum systemic inflammation and autoimmune diseases [5].

X-linked autoimmunity-allergic dysregulation syndrome (XLAAD), which has been renamed as Immunodysregulation, polyendocrinopathy, enteropathy, and X-linked syndrome (IPEX), was a rare inborn error of immune regulation and autoimmune lymphoproliferative illness in humans [6]. As one of the most well-known Mendelian disorders, IPEX is characterized by a loss of immunological tolerance caused by a lack of functioning Tregs and was discovered to be associated with the mutations in Foxp3 [7, 8]. In the absence of Tregs, activated $CD4^{+}$ T cells instigate multi-organ damage resulting in type 1 diabetes (T1D), enteropathy, eczema, hypothyroidism, and other autoimmune disorders.

Moreover, studies on Tregs depletion by cytotoxic T-lymphocyte antigen (CTLA)-4 Ab (e.g., ipilimumab) in tumor patients have shown a strong correlation between the induction of tumor regression and autoimmunity [9]. Ipilimumab acts not only on effector T cells (Teffs) but also on Tregs because the latter ones in both mouse systems and humans can be directly targeted by ipilimumab due to the constitutive expression of CTLA-4 on their cell surface [10, 11]. Except for a decreased frequency of circulating $CD25^{+}CD4^{+}$ Tregs can be observed upon ipilimumab, CTLA-4 blockade renders Teffs resistant to the inhibitory activity of Tregs, rather than modulating the immunosuppressive effects of Tregs on T cells and NK cells [12]. It is indicated that loss of Tregs has a close relationship with the development of autoimmune diseases from another perspective.

2.2 Immune tolerance and prevention of autoimmunity

Our body's immune system has evolved to perform self-tolerance to resist the autoimmune reactions directed against our own cells via sophisticated mechanisms. On the T cell level, self-tolerance is executed by deletion of T cells with self-reactive T cell receptor (TCR) in the thymus (central tolerance) or maintained by specialized cells, including Tregs, outside of the thymus (peripheral tolerance). The importance of Tregs for the maintenance of immune tolerance has also been illustrated to have a close relationship with the expression of the Foxp3 gene, both in humans and mice [13, 14]. Foxp3, together with other transcription factors and coactivators/corepressors, represses the transcription of IL-2 in Tregs, rendering them highly dependent on exogenous IL-2 (mainly produced by activated non-Tregs) for their maintenance and function. Tregs constitutively express the high-affinity IL-2 receptor (α chain), which serves as a sink for IL-2 that controls the expansion of Teffs. The development of autoimmune/inflammatory disease can be promoted if disrupting this IL-2-mediated feedback loop at any step. Further, manipulation of this feedback loop is instrumental in tuning the intensity of Tregs-mediated suppression, hence the strength of a variety of immune responses [15]. Foxp3 also activates the genes encoding Tregs-associated molecules, including CD25, CTLA-4, and Glucocorticoid induces tumor necrosis factor receptor (GITR) and confers suppressive activity to Tregs, which directly suppress non-Tregs or modulate the function of antigen-presenting cells (APCs) to activate non-Tregs [16].

2.3 Mechanisms of immune suppression

The Tregs-mediated immune suppression may be related to three mechanisms, including secretion of immunosuppressive cytokines [17], cell-contact-dependent suppression [18], and functional modification or killing of APCs [19]. More than one mechanism may operate for controlling the particular immune response in a synergistic and sequential manner.

IL-10 and TGF-β may act as the main immunosuppressive cytokines contributing to control the autoimmune disorders or inflammatory diseases secreted by Tregs [17]. IL-10 indirectly prevents antigen-specific T cell activation, which is associated with downregulation of the antigen presentation and accessory cell functions of monocytes, macrophages, and dendritic cells (DCs), as well as inhibits T-cell expansion by directly inhibiting IL-2 production by these cells. The pivotal function of TGF-β is to maintain tolerance via the regulation of lymphocyte proliferation, differentiation, and survival. TGF-β can block the proliferation of T lymphocytes by suppressing the expression of IL-2 (via Smad3 signaling pathway), cyclins (including cyclin D2 and cyclin E), cyclin-dependent kinase (CDK)-4, and c-myc. TGF-β also can inhibit the differentiation of Th1 and Th2 cells by blocking the T-bet/STAT4 and GATA-3/NFAT signaling transduction pathway, and down-regulating the differentiation of cytotoxic T lymphocyte (CTL) via regulating the expression of c-myc and T-bet [20]. TGF-β can also induce the expression of Foxp3 and the generation of Tregs. In addition, nTregs can also predominantly produce IL-35, a new member of the IL-12 family, to perform the suppressive function [21]. IL-35 is a novel Epstein-Barr-virus-induced gene (Ebi) 3-IL-12α heterodimeric cytokine, and Ebi3, which encodes IL-27 β, is a downstream target of Foxp3. $Ebi3^{-/-}$ and $IL12\alpha^{-/-}$ Tregs have significantly reduced regulatory activity *in vitro* and fail to control homeostatic proliferation and cure inflammatory bowel disease (IBD) *in vivo*.

Antigen-activated Tregs, which are highly mobile, are swiftly recruited to APCs (especially DCs), upon being stimulated by the specific antigen. The recruitment of Tregs to APCs is in chemokines or adhesion molecules depended on manner. Once the Tregs aggregate around the APCs, they will outcompete antigen-specific naïve T cells regarding interaction with DCs, mainly because of the high expression of adhesion molecules on Tregs, such as lymphocyte function-associated antigen (LFA)-1 [22].

Tregs can modulate the function of APCs. Activated Tregs promote the downregulation of CD80 and CD86 on APCs or stimulate DCs to form the enzyme indoleamine 2, 3-dioxygenase both by a CTLA-4-dependent mechanism [23, 24]. Indoleamine 2, 3-dioxygenase is capable of catabolizing the essential amino acid tryptophan to kynurenines, which are toxic to T cells. Alternatively, Tregs can induce the apoptosis of responder T cells or APCs by secreting granzyme/perforin or immunosuppressive cytokines (such as IL-10), or through the delivery of a negative signal (possible including intracellular cyclic AMP) to inactivate the responder T cells [19]. The upregulation of intracellular cyclic AMP will lead to the inhibition of T cell proliferation and IL-2 production, as well as the generation of pericellular adenosine catalyzed by CD39 and CD73 by Tregs.

3. Tregs and autoimmune and autoinflammatory diseases

3.1 Tregs and type 1 diabetes

Type 1 diabetes (T1D) is a typical kind of autoimmune disease affecting millions of people worldwide with a steadily rising incidence, and islet infiltrating self-reactive T cells mediated β-cell destruction is considered to be primary pathogenesis of this disease. The initiation of the autoimmune process is related to the recognition of self-antigens by the autoreactive subsets of $CD4^+$ T-helper lymphocytes, which can preferentially produce the Th1 cytokine spectrum after activation. The presence of autoreactive $CD8^+$ cytotoxic T lymphocytes is necessary for the further development of T1D as well. It has been demonstrated that $CD4^+CD25^+FoxP3^+$ Tregs also play an indispensable role in the development of T1D by preventing destructive autoimmunity [25]. Although the application of immunosuppressive reagents is one of the available therapies, it can have severe side effects. Optimal immune-based therapies for T1D should restore self-tolerance without inducing chronic immunosuppression. Thus, efforts to repair or replace Tregs in T1D probably can reverse autoimmune response and protect the remaining insulin-producing β cells. There is a large body of evidence to suggest that $Foxp3^+$ Tregs function is altered in patients with T1D, though the overall frequency of $Foxp3^+$ Tregs may be unaltered in these individuals [26, 27]. Data from the non-obese diabetic (NOD) mouse model of autoimmune diabetes and human with T1D suggest that increasing resistance of Teffs to Tregs regulation may be the primary cause for reduced suppression, and it can be explained by the inability of Teffs to provide an environment conducive to Tregs fitness and function, including the reduced IL-2 production or downregulation of the IL-2 signaling pathway by Teffs [28, 29].

Apart from their canonical function of immune suppression, it is now well accepted that Tregs can likewise be induced in the periphery in an antigen-specific manner and take residence in tissues to play important roles in maintaining tissue homeostasis. So, antigen-specific induction of disease-relevant Tregs will offer the opportunity to treat or prevent the T1D for a long-standing goal. It has been demonstrated that in the peripheral blood of children who are at risk to develop T1D, the proportion of insulin-specific Tregs reduced during the onset of islet autoimmunity, while the higher reduction was related to a rapid progression to clinically overt T1D [30]. This finding suggested that inducing these insulin-specific Tregs may delay the progression to clinically symptomatic T1D. Nevertheless, very little is known about pancreas residing Tregs, and all studies conducted so far on these tissue-specific Tregs focused solely on NOD mice with ongoing insulitis. A recent study found that a combinatorial regimen involving the anti-CD3, cyclophosphamide (CyP), and IAC (IL-2/JES6–1) antibody complex can promote the engraftment of antigen-specific donor Tregs through ablating host conditioning and control islet autoimmunity without long-term immunosuppression [31].

3.2 Tregs and rheumatoid arthritis

Rheumatoid arthritis (RA) is one kind of common systemic inflammatory autoimmune disease, and its typical clinical symptoms are musculoskeletal pain, joint swelling, and stiffness, which can seriously damage body function and reduce the quality of life of patients. Patients with RA are more likely to develop osteoporosis, infection, cardiovascular diseases, respiratory diseases, cancer, and other diseases than the general population. Similar to other autoimmune diseases, Tregs also play a vital role in the pathogenetic process in RA. When the number and/or function of Tregs are decreased or inhibited, autoantigen or ligand death receptors (DRs) related immune cascade can be amplified, and the levels of various cytokines, such as IL-2, will be rapidly increased, leading to the activation of macrophages in the synovium of bones and joints to produce many inflammatory cytokines including IL-1, IL-6, and IL-8 [32, 33]. These inflammatory reactions destroy articular cartilage and eventually lead to joint deformities.

However, contradictory results on the number (increased [34], unchanged [35], or decreased [36]) and functional characteristics (enhancement [32] or attenuation [33]) of Tregs in the peripheral blood of patients with RA have been reported in different studies, and this discrepancy can be explained by the ongoing difficulties in the recognition of Tregs. In most studies, the high-level expression of Foxp3, CD25, and low-level expression of CD127 (the α-chain of the IL-7 receptor) are used to define Tregs, and the $CD3^{+}CD4^{+}CD25^{high}CD127^{low}$ phenotype is most commonly isolated from Tregs population. However, Foxp3 requires intracellular staining and the expression levels in Tregs in the resting state and activated state are different, and conventional T cells (Tconvs) also express a low levels of Foxp3 and CD25 upon TCR stimulation and low levels of CD127 [37]. Thus, some other supplementary cell surface markers, such as CD62 ligand, integrin Ea (CD103), GITR (TNFRSF18), CTLA-4 (CD152), CD45RO, and neuropilin, have been also used to identify Tregs in clinical practice [38]. CD45RA and CD45RO can be used to distinguish immature Tregs ($CD45RA^{+}Foxp3^{low}$) from activated memory Tregs ($CD45RA^{-}Foxp3^{high}$) cells [39]. A more stringent method to define Tregs has revealed the number of Tregs decreased in peripheral blood and increased in synovial fluid by performing a meta-analysis [40].

However, although the Tregs isolated from RA patients can show normal inhibitory activity *in vitro*, they function abnormally when circulating in the synovial fluid, which is caused by the overexpression of IL-6 induced inflammatory environment [41, 42]. Teffs in this inflammatory environment are resistant to Tregs-mediated repression, and the sensitivity of APCs to the inhibition of Tregs also decreased [43, 44]. Moreover, the arthritic synovial fibroblasts can promote the transformation of $CD25^{low}Foxp3^{+}CD4^{+}$ T cells into Th17 cells in the oxygen deficiency synovial microenvironment, and the latter one shows a stronger ability to induce osteoclast production [45]. The process of transformation is closely related to the activation of the hypoxia-inducible factor-1a (HIF-1a) pathway.

3.3 Tregs and autoimmune hepatitis

Autoimmune hepatitis (AIH) is a severe hepatopathy that occurs globally in all ethnicities and affects children and adults of all ages. It is with a female predominance and characterized by hypergammaglobulinemia, interface hepatitis on histology, and seropositivity for disease-defining autoantibodies. In AIH, the autoimmune reaction resulting in liver injury initiates with the presentation of liver autoantigen by APCs to an uncommitted T lymphocyte. Following antigen encounter, Th0 becomes activated and differentiates into Th1, Th2, and Th17. Th1 cells secrete interferon (IFN)-γ and IL-2, which can lead to the activation of macrophages and upregulation of major histocompatibility complex (MHC) class I and II by hepatocytes [46]. Th2 lymphocytes secrete IL-4 and IL-10, which can promote the B cell activation and maturation into plasma cells. Plasma cells then produce autoantibodies and mediate cell cytotoxicity in turn [47]. Activation of Th17 cells, which can secrete IL-17 proinflammatory cytokines, has been associated with the induction of pro-fibrotic events [48]. The autoimmune attack will continue perpetrating and favoring the progression of tissue damage if these events are not opposed by effective immunoregulation.

It has been demonstrated that the impairment of Tregs plays an important role in the initiation and progression of AIH. A numerical and functional defect in $CD4^{+}CD25^{+/high}Foxp3^{+}$ cells was reported in patients with AIH compared with the healthy subjects [49]. Before immunosuppressive treatment is instituted, Tregs isolated from AIH patients are also impaired in their ability to expand, and unable to regulate $CD4^{+}$ and $CD8^{+}$ T cells proliferation and modify $CD4^{+}$ and $CD8^{+}$ T cells cytokine profile as in the case of healthy controls [50, 51].

This deficiency of Tregs in AIH patients might be linked to increased expression of the cell surface marker CD127 [52] and defects in the expression of the ectonucleotidase CD39 [53]. CD127 is also known as the α-chain of the IL-7 receptor (IL-7Rα), while $CD39^{+}$ Tregs decrease in frequency in AIH patients leading to the failure to control the production of IL-17 by Th17. So, Tregs in AIH subjects are more prone to acquire features of effectors than their counterparts when exposed to a proinflammatory challenge, which suggests the defective immunoregulation of Tregs in AIH might have some relationship with the increased conversion of Tregs into effector lymphocytes [53]. A recent study confirmed that impaired CD39 levels derive from alterations of aryl hydrocarbon receptor (AhR) signaling [54]. AhR is a mediator of toxin responses and adaptive immunity. Upon binding to endogenous or exogenous ligands, AhR undergoes activation, which will bring about the upregulation of CD39.

However, aberrantly high levels of aryl hydrocarbon receptor repressor and estrogen receptor alpha (Erα) can be detected in AIH. AhR binds Erα with higher affinity than aryl hydrocarbon receptor nuclear translocator (ARNT), the classical AhR binding partner. These non-conventional binding give rise to impaired CD39 upregulation.

Impaired Tregs function in AIH is also linked to defective levels of Galectin-9 (Gal-9). Gal-9, a member of the galectin family, is one kind of β galactoside binding protein expressed on Tregs. It can bind to the mucin domain 3 (Tim-3) on $CD4^{+}CD25^{-}$ Teffs. Upon Gal-9 binding to Tim-3, apoptosis in $CD4^{+}CD25^{-}$ Teffs will be induced. Thus, reduced expression of Gal-9 in Tregs in AIH contributes to the less suppressing ability of Tregs and rendering $CD4^{+}CD25^{-}$ Teffs less prone to the control of Tregs [55].

In addition, defective Tregs function in AIH is linked with reduced ability to produce IL-10 as well. It is resulting from poor response to IL-2 as reflected by impaired ability to upregulate the phosphor signal transducer and activator of transcription 5 (pSTAT-5) [56].

3.4 Tregs and inflammatory bowel disease

Inflammatory bowel disease (IBD) is a chronic, inflammatory, and autoimmune disorder. The types of IBD include ulcerative colitis (UC) and Crohn's disease. The etiology of IBD is possibly linked to the dissonance of the host immune system, genetic variability as well as an environmental factor, and the pathogenesis of this disorder has not been fully elucidated [57]. In recent years, it has been found that the abnormal intestinal mucosal immune system plays a crucial role in the occurrence, development, and prognosis of IBD, involving the imbalance in Th17 and Tregs [58]. The differentiation of Th17 cells goes through three stages—initiated by IL-6 and TGF-β, expanded by IL-21, and IL-23 maintains the stable maturation of Th17 cells during the later stage of differentiation [59]. Except for protecting the intestinal mucosa via keeping the balance of the immune microenvironment, Th17 cells also can exacerbate the intestinal inflammatory response through secreting proinflammatory cytokines, such as IL-17. Compared with the healthy controls, Th17 cells infiltrate the intestinal mucosa of IBD patients and the amount of IL-17 increases [60]. Tregs and Th17 cells are related through differentiation and share a common signal pathway mediated by TGF-β. In the UC mouse model, Th17 cells in the peripheral blood of mice increased, yet Tregs decreased [61]. Therefore, Tregs deficiency may be the central link in the pathogenesis of IBD and the regulation of Th17/Tregs balance is prospective to be a new target for the treatment of IBD. The immunological factors affecting the Th17/Tregs balance in IBD consist of both TCR and costimulatory signals and cytokines. IL-2 inducible T cell kinase (ITK), a critical regulator of intracellular signaling downstream of the TCR, positively regulates the differentiation of Th17 and negatively regulates the differentiation of Tregs [62]. The T cell costimulatory molecule OX40 and its cognate ligand OX40L collectively play an essential role in keeping the growth of Th17 and Tregs, that is, activation of OX40 enhanced Th17 function while blocking OX40L decreased Tregs proliferation [63].

4. Tregs and transplantation

While organ transplantation is one of the greatest achievements in modern medicine, rejection is still the major barrier to successful transplantation. The immune response to an allograft is an ongoing dialog between the innate and adaptive immune systems. One of the reasons that transplantation induces such a dynamic immune response is the high precursor frequency of T cells capable of responding to mismatched MHC molecules. Although immunosuppression regimens are effectively able to control the acute rejection and decrease graft loss in the first year after transplantation, it is difficult to get a durable effect on long-term graft survival with these modern regimens, owing to a combination of drug toxicities, the emergence of chronic alloimmune responses and the serious complications, such as chronic infections or malignancies. Studies on experimental transplant models have suggested a role for Tregs in protecting allografts by suppressing both autoimmune and alloimmune responses [64, 65]. Further, Tregs-based therapies do not require harsh conditioning and have a risk of graft-versus-host disease.

4.1 Tregs and solid organ transplantation

The first step in the adaptive immune response to a transplant in a solid organ transplantation recipient is T-cell recognition of alloantigen or allorecognition. Graft-specific Tconvs, which are capable of direct recognition of alloantigen, are present at a very high frequency so that they can respond to the transplant without first clonally expanding in lymph nodes. When graft-specific Tconvs are recruited to the graft, they will lead to inflammation and tissue damage. Increasing graft-specific Tregs combined with the reduction of graft-specific Tconvs allow the former one to dominate in the graft and prevent recruitment and activation of the later one. Moreover, once entering the draining lymph node, inflammatory APCs can activate more graft-specific Tconvs, while tolerogenic APCs are able to expand graft-specific Tregs and prevent the expansion of graft-specific Tconvs to maintain tolerance [66].

Moreover, Tregs with direct alloantigen specificity, which are also present at high frequency, play important role in the induction of tolerance, whereas Tregs with indirect alloantigen specificity are important for the maintenance of tolerance [67]. Tregs control transplant rejection by first migrating to the organ to prevent graft damage and then retreating to draining lymph nodes to maintain tolerance [65]. During an active alloimmune response, Tregs with both direct and indirect specificities expand and infiltrate the organs, but the homeostatic function of Tregs is insufficient to prevent rejection from occurring due to the potency of alloimmune responses until the organs have suffered substantial damage [67]. That is because Teffs arrive at the graft site first and expand in number before the arrival of Tregs so that the grafts are dominated by Teffs [64], and at the peak of alloimmune responses, a high antigen load, vigorous co-stimulation, and high concentrations of cytokines, such as IL-1 and IL-6, override Tregs suppression so that effective immune functions can be carried out to induce rejection quickly. Thus, prevention of rejection and establishment of tolerance by Tregs require attenuation of Teffs responses and inflammation control.

4.2 Tregs and allogeneic hematopoietic cell and bone marrow transplantation

Currently, allogeneic hematopoietic cells transplantation (HCT) or bone marrow transplantation (BMT) in humans is widely used in the treatment of tumors of the hematopoietic and immune systems, including leukemia, lymphoma, and myeloma. However, they are usually complicated by serious and potentially lethal side effects, such as immunodeficiency and graft-versus-host disease (GVHD). GVHD represents a dysregulated immune response and has been assessed across both major and minor histocompatibility barriers, and the pace of these reactions is much more accelerated across major histocompatibility barriers. The onset and course of GVHD depend on the degree of major and minor MHC disparity and the T-cell dose. It has been demonstrated by using animal models that T cells rapidly migrate to nodal sites, spleen, and mesenteric lymph nodes and begin to dramatically expand by 3–4 days following adoptive transfer, and within 7–10 days, they infiltrate the major sites of GVHD pathophysiology, such as lymph nodes, spleen, gastric intestinal tract (GI tract), liver, and skin [68]. Depletion of $CD4^+CD25^+$ T cells from the donor graft accelerated the GVHD course and increased lethality, which provided evidence for the role of Tregs in mediating GVHD [69]. Tregs also expand dramatically upon adoptive transfer and traffic to nodal sites to promote immune reconstitution and suppress GVHD across both major and minor histocompatibility barriers, while interestingly allowing for the maintenance of graft-versus-tumor (GVT) responses [70–72]. Tregs proliferate in the same way as Tconvs with similar kinetics and tend to fade out over time. Upon the adoptive transfer, the dramatic expansion of Tconvs can be detected, whereas when the same numbers of Tregs were adopted along with the Tconvs, this dramatic proliferation of Tconvs is significantly reduced, yet the homing and activation of Tconvs are not impacted [73]. It is indicated that the adopted Tconvs are still able to be activated and home to specific sites within the body, yet this drastic T-cell expansion required for GVHD is diminished. Thus, clinical strategies to enhance the function of Tregs hold great promise to improve outcomes following allogeneic HCT and BMT.

5. Tregs therapy

The ability of Tregs to maintain self-tolerance means they are critical for the control and prevention of autoimmune diseases. Currently, a large body of data in the literature has provided evidence on the possible Tregs therapy for various immune-mediated diseases. Restoring immune homeostasis and tolerance through the promotion, activation, or delivery of Tregs has emerged as a focus for therapies aimed at curing or controlling autoimmune diseases. A variety of Tregs-based therapies are being explored in the treatment and prevention of autoimmune diseases, such as *ex vivo*-expanded polyclonal Tregs or Tregs transduced with an autoantigen-specific TCR, chimeric antigen receptor (CAR). In addition, some other non-cell-based therapies related to Tregs, including low-dose IL-2 and heat shock protein (HSP), may also be beneficial.

5.1 Polyclonal Tregs therapy

Polyclonal Tregs therapy uses autologous *ex vivo*-expanded Tregs to restore tolerance and is considered a next-generation cellular therapy for several autoimmune diseases and inflammatory immune disorders. Tregs isolated from peripheral blood are stimulated and expanded *in vitro* by using anti-CD3/CD28 antibody-coated beads and high dose IL-2, or anti-CD28 super agonists [74, 75]. The first preclinical proof of concept for use of polyclonal Tregs was demonstrated in 1995 that $CD4^+CD25^+$ T cells could be used to transfer tolerance in athymic nude mice by suppressing self-reactive lymphocytes [76]. Then, expanded Tregs with a polyclonal specificity are reportedly more efficient in suppressive function and have demonstrated potential in various preclinical models of GVHD [77], solid organ transplantation [78], and autoimmune diseases [79]. Several clinical trials have been carried out to examine the safety and feasibility of polyclonal Tregs for T1D [80], transplantation [81], and GVHD [82], and the use of polyclonal Tregs in these diseases have shown significant therapeutic potential. For example, one robust clinical trial on T1D with polyclonal Tregs demonstrated that the expanded autologous Tregs retained their T cell receptor diversity and owned enhanced functional activity. Fourteen adult subjects with T1D received *ex vivo*-expanded autologous $CD4^+CD127^{low/-}CD25^+$ polyclonal Tregs (0.05×10^8 to 26×10^8 cells). A transient increase in Tregs, which retained a broad $CD4^+Foxp3^+CD25^{hi}CD127^{low}$ phenotype long-term, was detected in recipients. There were no infusion reactions or cell therapy-related high-grade adverse events [80]. Besides, some other clinical trials for polyclonal Tregs therapy in autoimmune hepatitis (NCT02704338), Crohn's disease (NCT03185000), Pemphigus (NCT03239470), and Alzheimer's disease (NCT03865017) are also under investigation. Positive results from these clinical trials have allayed the concerns that polyclonal Tregs therapy would promote generalized immune suppression, leading to an increased risk of infection and cancer, which has not been found.

However, the therapeutic effect of this clinical trial correlated with increased Tregs post-infusion, and only persisted for a short time. The subsequent trials confirmed the limited persistence of expanded Tregs even after a second infusion [83], and obtaining sufficient cell numbers can be challenging in many disease scenarios [84], although polyclonal Tregs therapy is generally considered safe and efficacious. Perhaps the use of other Tregs-promoting therapies in combination with polyclonal Tregs therapy would prolong the suppressive effect and increase the number of Tregs with improved patient outcomes.

5.1.1 Low-dose IL-2 in combination with polyclonal Tregs therapy

As mentioned earlier, it has been widely accepted that IL-2 plays a critical role mainly in Tregs fitness and homeostasis, thus low-dose IL-2 therapy alone has the effect of expanding *in vivo* Tregs. Co-administration of polyclonal Tregs and low-dose IL-2 has been considered as an additional strategy to restore the defective Treg pool and is expected to boost Tregs number and function after administration. Recent work has reported the possibility to expand Tregs using low-dose IL-2 *in vivo* [85]. In this phase I-II clinical trial on 46 individuals with mild to moderate forms of various autoimmune diseases, including RA, ankylosing spondylitis, systemic lupus erythematosus, psoriasis, Bechet's disease, granulomatosis with polyangiitis, Takayasu's disease, IBD, AIH and sclerosing cholangitis, all the patients received low-dose IL-2 (1 million IU/day) for 5 days, followed by fortnightly injections for 6 months. Low-dose IL-2 can be well tolerated whatever the disease and the concomitant treatments, and specific Tregs expansion and activation can be detected in all patients, without effector T cell activation. The increase in Tregs percentage was mainly evident on day 8 and was then contained thereafter, despite the levels remaining slightly higher compared to baseline. However, specific data on AIH patients ($n = 2$) were not presented, which indicated the effects of low-dose IL-2 in AIH patients remain unclear. In another study on two AIH patients with persistent disease activity [86], low-dose IL-2 was administered at 1 million IU for 5 days monthly for a total of 6 months. The proportion of circulating Tregs increased in both cases with a peak observed on day 9 and returning to baseline levels on day 28. This suggests that the effect of low-dose IL-2 on Tregs frequency is transient. Given the small number of cases enrolled, further studies in larger numbers of subjects should be performed to assess the efficacy as well as the long-term effects of this treatment on Tregs, particularly on their suppressive function, expansion, and plasticity.

Furthermore, an advantage of low-dose IL-2 therapy is that recombinant human IL-2 is already available as a therapeutic drug called Aldesleukin or Proleukin for the treatment of malignant melanoma and renal cell carcinoma in the clinic [87].

5.1.2 HSP in integrating Tregs expansion and activation

HSPs are highly conserved proteins present in all kingdoms of organisms, and expressed under stress conditions to protect the cells from injuries. They are classified into five families according to their molecular weight, including HSPH, HSPC, HSPA, HSPD, and DNAJ. Intracellular HSPs play an essential role in physiological processes, involving of folding of nascent and stress-accumulated protein-substrate assembly and preventing the aggregation of these proteins, transporting across membranes and degrading other proteins. While extracellular or receptor-bound HSPs mediate immunological functions and immunomodulatory activity, including the induction, proliferation, suppressive function, and cytokine production of Tregs [88]. In patients with juvenile idiopathic arthritis (JIA), DNAJ was found to improve the suppressive function of Tregs in culture and stimulate T cells for the production of IL-10, and high serum levels of DNAJ correspond with a milder course of the disease, indicating epitopes derived from human DNAJ can induce differentiation and/or stimulate cell proliferation of Tregs [89]. Acting as co-stimulators of human Tregs, HSPD can enhance the suppression and proliferation of Tregs via binding of Toll-like receptor (TLR) 2 on the Tregs surface to inhibit target T cell proliferation, IFN-γ and tumor necrosis factor (TNF)-α secretion, as well as upregulate the expression of IL-10 [90]. HSPD can enhance the differentiation of cord blood mononuclear cell (CBMC) into $CD4^{+}IL\text{-}10^{+}Foxp3^{+}$ Tregs as well [91]. HSPA can stimulate the suppressive activity of Tregs, increase the production of IL-10, and downregulate the production of inflammatory cytokines via the TLR4-signaling pathway, which may be important for Foxp3 induction [92]. Animal studies have shown that oral, nasal, intraperitoneal, or intradermal administration of HSPA significantly inhibits the development of the autoimmune arthritic model, which suggested that suppression of autoimmune response in experimental animals was mediated by increased expansion of Tregs specific for HSPA, and the secretion of anti-inflammatory IL-10 [93–95]. Moreover, HSPC can promote Tregs-dependent suppression as well [96]. HSP gp96, the endoplasmic reticulum form of HSPC, is required for Tregs maintenance and function, as loss of GP96 resulted in instability of the Tregs lineage and impairment of suppressive functions *in vivo*. In the absence of HSP gp96, Tregs are unable to maintain Foxp3 expression levels and can lead to systemic accumulation of IFN-γ-producing and IL-17-producing T cells, because HSP gp96 is an essential chaperone for the cell-surface protein glycoprotein A repetitions predominant (GARP). GAPR is a docking receptor for latent membrane-associated TGF-β (mLTGF-β). The HSP gp96-deficient Tregs prevent the expression of mLTGF-β and resulted in inefficient production of active TGF-β [97]. Meanwhile, immunization of HSP gp96 can increase Tregs frequency, expansion, and suppressive function, which shows obvious therapeutic effects in a $Lyn^{-/-}$ mouse model of systemic lupus erythematosus and myelin oligodendrocyte glycoprotein (MOG)-induced experimental autoimmune encephalomyelitis (EAE) [98].

5.1.3 Foxp3-transduced T cells

Tregs constitutively express the transcription factor Foxp3, which is critical for their immunosuppressive function. Several studies have provided evidence that ectopic expression of Foxp3 can confer a suppressive phenotype to naïve or memory $CD4^{+}$ T cells, so it is probably a way to circumvent the requirement of a large number of polyclonal Tregs for therapy [99]. Lentiviral delivery of the Foxp3 gene into IPEX patient-derived $CD4^{+}$ T cells can acquire the characteristic features, such as decreased proliferation, hyporesponsiveness, reduced cytokine release, and suppressive activity, which are able to mirror the Tregs population from healthy donors, and these induced Tregs were demonstrated to be stable in inflammatory conditions not only *in vitro* but also *in vivo* in a xenograft mouse model of GVHD [100]. Several other studies have also shown the efficiency of Foxp3-transduced Tregs in combating autoimmune diseases, such as allergy [101] and collagen-induced arthritis [102]. Recently, the CRISPR/Cas9 system has been applied to the domain for stable and high-level expression of Foxp3 in Tconvs, and these edited Tregs-like cells were able to suppress the immune response in a xeno-GVHD mouse model [103]. These studies demonstrate the applicability of gene correction in the treatment of autoimmune diseases.

Cell permeable form of Foxp3 is another approach to enforce Tregs differentiation. This protein form can link to the protein transduction domain (PTD) from the HIV transactivator of transcription and allow Foxp3 to be delivered to the cytoplasm and nucleus, which has been shown to induce a Tregs phenotype in both human and mouse T cells [104]. However, a major limitation of this approach is the high cost for human patients.

5.2 Antigen-specific Tregs

While the initially limited success of polyclonal Tregs is encouraging, the amounts of cells needed for infusions are quite large and the risk of nonspecific immunosuppression should be considered. Tregs developed in the thymus (i.e., nTregs) harbor a TCR repertoire that is skewed toward self-antigens, while Tregs induced in the periphery in an antigen-specific manner (i.e., pTregs) can be characterized with a TCR repertoire different from their nTregs counterparts [105]. So, it is a good strategy to induce disease-relevant antigen-specific Tregs with the goal to interfere with the unwanted immune reactions in allergies and autoimmunity and to restore the self-tolerance, and it has been verified through considerate research in humans or mice [106, 107]. Compared with the polyclonal Tregs therapy, growing evidence from animal models indicates that antigen-specific Tregs may be more efficient in controlling pathological immune responses in a disease-specific manner. It is possible because infused Tregs migrating toward the tissues of cognate antigen exposure will lead to more effective and localized control of inflammation, along with risk reduction of broad immunosuppression and its related adverse events [108, 109]. Moreover, the enhanced migration ability of antigen-specific Tregs to target tissues can probably lead to a lower administration number of Tregs than the polyclonal approach, and facilitate the obtainment of Tregs via standard *in vitro* expansion protocols.

5.2.1 TCR-Tregs therapy

Tregs therapy can be enhanced by the introduction of an autoantigen-specific TCR (TCR-Tregs), which have the ability to redirect their response toward the desired autoantigen specificity. Tregs can be *ex vivo*-transduced to express a high-affinity and autoantigen-specific TCR by way of retroviral or lentiviral transduction and subsequently expanded to treat a specific autoimmune disease. For instance, a low number of autoantigen-specific Tregs were needed to sufficiently prevent or even reverse T1D in a NOD mice model, which was engineered to express a diabetogenic TCR [108]. Another group showed that as few as 2000 antigen-specific Tregs were all that was required to prevent T1D in mice [110]. Preclinical studies in mouse models have also shown that TCR-engineered Tregs are more effective in suppressing the Teffs responses against specific antigens in autoimmune diseases, such as colitis, multiple sclerosis, and arthritis [111–113], even in tolerance induction to MHC mismatched heart grafts [114].

Compared to polyclonal Tregs, fewer antigen-specific Tregs may be needed to alleviate autoimmune disease; however, the challenge of the identification of an appropriate, high-affinity, autoantigen-specific TCR for transduction onto Tregs still remains, due to some autoimmune diseases being with poorly defined dominant epitopes. It is hard to isolate and identify antigen-specific Tregs due to both the great diversity in TCRs and very low count of them naturally circulating in the peripheral blood. The majority of antigen-specific Tregs were generated using TCRs isolated from Tconvs, which would influence the stability, avidity, and migration to specific parts of the engineered Tregs, for the reason that the intrinsic affinity and specificity of TCRs isolated from Tregs are distinct from Tconvs. Moreover, there are some other limitations of this approach, such as the requirement for MHC restriction and the risk of mispairing with endogenous TCR.

Single-cell sequencing is required for TCR identification since each T cell clone expresses a different TCR sequence from the others, and the successful sequencing of both the α and β chain TCR is required to successfully identify one TCR [115]. In a recent study, single-cell TCR analyses of islet Tregs revealed their specificity for insulin and other islet derived antigen, and these antigen-specific Tregs were reported to be efficient in protecting NOD mice from diabetes [116].

5.2.2 CAR-Tregs therapy

Although Tregs engineered with TCRs (TCR-Tregs) seem to be promising, they are still MHC-restricted and their modular application in individual patients is constrained. Engineer with genes encoding chimeric antigen receptors (CARs), which typically consist of a single-chain variable fragment (scFv) for binding to a monoclonal antibody, an extracellular hinge, a transmembrane region, and intracellular signaling domains, is an MHC-independent strategy of generating antigen specificity for Tregs [117]. In animal models, CAR-Tregs have shown great potential for treating different diseases, especially allograft rejection and various autoimmune diseases.

HLA-A mismatching is often associated with poor outcomes after transplantation, so, HLA-A is a potential target antigen to generate antigen-specific Tregs for inducing transplantation tolerance. One kind of HLA-A2-specific CAR (A2-CAR) Tregs was created in a peptide-independent manner, and not only can maintain high expression of canonical Tregs markers, including Foxp3, CD25, Helios, CTLA-4, and a high degree of demethylation of the Treg-specific demethylated region (TSDR) of the FOXP3 locus but also can enable stronger antigen-specific activation than did an endogenous TCR [118]. Further, CAR-stimulated Tregs had a higher surface expression of CTLA-4, latency-associated peptide (LAP), and the inactive precursor of TGF-β than TCR-Tregs. Unlike TCR-Tregs, CARs could also stimulate IL-2-independent Tregs proliferation in the short term [118]. Thus, CAR-Tregs may be superior to TCR-Tregs.

CAR Tregs isolated from transgenic BALB/c mice with a CAR specific for 2,4,6-trinitrophenol (TNP), an antigen commonly used in a mouse model of colitis, were reported to capable of suppressing the proliferation of Teffs *in vitro* even in the absence of B7-CD28 co-stimulation, and the mortality rate of TNP-CAR-tg mice significantly decreased in comparison with WT mice [119]. In situ fluorescent micro endoscopic evaluation verified that TNP-CAR Tregs localized to the inflamed colonic mucosa. Thereafter, a novel protocol that enabled efficient and reproducible retroviral transduction and expansion of murine nTregs was developed in a non-transgenic mouse model, bringing about a highly enriched population of TNP-specific Tregs. The TNP-CAR Tregs show suppressive capabilities to Teffs both *in vitro* and *in vivo*, and TNP-CAR Tregs-mediated suppression *in vitro* was partially dependent on cell-cell contact but not on IL-10 or TGF-β1 [120]. Moreover, based on the previously engineered Tregs to express a TCR specific for a myelin basic protein (MBP) peptide, which can suppress the proliferation of MBP-reactive Teffs and ameliorated MOG-induced EAE, the approach by creating human Tregs expressing functional single-chain CAR (scFv CAR), targeting either MBP or MOG was extended. These scFv CAR-transduced Tregs retained Foxp3 and Helios after long-term expansion *in vitro*. Importantly, these engineered CAR-Tregs were able to suppress autoimmune pathology in EAE, demonstrating that these Tregs have the potential to be used as a cellular therapy for multiple sclerosis (MS) patients [121].

5.3 Taking Tregs into medicine

5.3.1 Comparison table showing different approaches, techniques, and stages among studies

Different approaches that involve boosting Tregs have been tested in several disease settings so for. Polyclonal Tregs and antigen-specific Tregs therapy have demonstrated their efficacy in immunotherapy in various clinical trials or preclinical models (**Table 1**).

5.3.2 Challenge and bottleneck of Tregs therapy

To sum up, Tregs are crucial in maintaining tolerance. Hence, Tregs immunotherapy is an attractive therapeutic option in autoimmune diseases and organ transplantations. However, there are still many challenges and bottlenecks in implementing Tregs therapy.

At first, the cellular variability of Tregs is wide. It is important to characterize the phenotype and suppressor function of each subtype of Tregs present in the periphery or the thymus. The success of Tregs therapy depends initially on the isolation and characterization of cells, while current research does not use a universally applicable standard for Tregs identification. This gap in identification leads to conflicting and doubtful research results. Meanwhile, one of the drawbacks of this cell therapy is the time delay to administer Tregs from taking peripheral blood to obtaining sufficient numbers of cells, and antigen-specific Tregs technology may presumably need administration of lower Tregs numbers than polyclonal approachesSecondly, to improve the efficacy of Tregs immunotherapy, it is necessary that Tregs can migrate, survive, and function in the specific target tissue. The plasticity of polyclonal or CAR-Tregs in an inflamed microenvironment is still an unknown factor. The inflamed microenvironment enriched with pro-inflammatory cytokines can either lead to a reduction in the potency of Tregs or resistance of Teffs to Tregs suppression, or even converting Tregs into pathogenic Teffs. There are also questions to be addressed regarding the long-term proliferative potential and survival of polyclonal or antigen-specific Tregs in the tissue microenvironment, which is enriched with cytokines, metabolites, low oxygen levels, and microbial peptides.

Thirdly, the application of CAR-Tregs is an exciting option in both transplantation and autoimmune diseases, when the antigen is known. Nevertheless, before CAR-Tregs can be put into practice in the clinic, there are still obstacles required to be overcome, because antibodies specific for self- or alloantigen must be characterized to construct antigen-specific CAR-Tregs. For the reason that autoimmune diseases always have a large autoantigenic repertoire of T or B cells, or spreading epitope, it will be not adequate to focus Tregs therapy on one specific epitope for an autoantigen. Although CAR-Tregs own a greater affinity to the cognate antigen than TCR-Tregs,

Approach	Technique	Indication	Stage of study	Study ID or references
Polyclonal Tregs therapy	Autologous polyclonally expanded Tregs	T1D	Clinical trials phase 1 (completed)	NCT01210664
Polyclonal Tregs therapy	*Ex-vivo* expanded donor regulatory T cells	GVHD	Clinical trials phase I (active)	NCT01795573
Polyclonal Tregs therapy	Autologous polyclonally expanded Tregs	Kidney transplant	Clinical trials phase I/II (Active)	NCT02129881
Polyclonal Tregs therapy	Donor alloantigen reactive Tregs	Liver transplant	Clinical trials phase I (recruiting)	NCT02188719
Polyclonal Tregs therapy	Autologous polyclonal expanded nTregs	AIH	Clinical trials phase I/II (unknown)	NCT02704338
Antigen-specific Tregs therapy	$CD4^+CD25^+$ T cells from TCR-transgenic BDC2.5 mice expanded *in vitro* with BDC peptide and NOD DCs	T1D	Preclinical studies (NOD model)	[122]
Antigen-specific Tregs therapy	$CD4^+$ T cells transduced with Foxp3 and a TCR of a CIA-associated T cell clone	RA	Preclinical studies (DBA1 mice)	[123]
Antigen-specific Tregs therapy	$CD4^+CD25^+$ T cells from TCR-transgenic Tg4 mice expanded *in vitro* with anti-CD3/CD28 beads	MS	Preclinical studies (B10.Pl mice)	[124]
Antigen-specific Tregs therapy	CAR-engineered $CD4^+CD25^+$ Tregs specific for CEA	Colitis	Preclinical studies (CEABAC mice)	[111]
Antigen-specific Tregs therapy	CAR-engineered human $CD4^+CD25^+$ Tregs specific for HLA-A2	Skin transplantation	Preclinical studies (CEABAC mice)	[125]
Antigen-specific Tregs therapy	TGF-β-induced iTregs generated from $CD4^+$ T cells of TxA23 mice	Autoimmune gastritis	Preclinical studies (BALB/c mice)	[126]
Antigen-specific Tregs therapy	TGF-β-induced OVA-specific iTregs generated from $CD4^+$ T cells of OT-II mice	GVHD	Preclinical studies (C57Bl/6 mice)	[127]

Table 1.
Clinical trials or preclinical models with polyclonal Tregs or antigen-specific Tregs in different diseases.

the former requires the target cells to have at least 100 target autoantigens for successful recognition and Tregs stimulation. Moreover, it has not been confirmed yet whether CAR-Tregs would also lead to adverse reactions, such as cytokine storm and neuronal cytotoxicity, as the treatments with anti-tumor CAR-T cells.

Besides, it is still a hard nut to crack to access the localization of infused Tregs to the exact target site, and exhaustion of Tregs may limit their efficacy in immunosuppression. Meanwhile, the choice of immunosuppression in patients with Tregs therapy is crucial, for example, rapamycin has been shown to enhance Tregs frequency. To achieve efficacious and successful Tregs therapy, it is necessary to continue on immunosuppression that is favorable to Tregs survival and proliferation.

Therefore, more work is required to administer Tregs therapy effectively and safely to restore tolerance in transplantations and autoimmune diseases.

6. Prospect (perils and promises)

Tregs have proved to be a major breakthrough as an exciting immunotherapy option in the last two decades. Early phase clinical trials demonstrated safety, feasibility, and early efficacy with Tregs therapy in both autoimmune diseases and organ transplantation. The development of antigen-specific Tregs and CAR-Tregs would lead to exciting new frontiers in the cell therapy field as these cells are more efficacious and lesser numbers are required due to their target tissue homing affinity. It is crucial to obtain tissue biopsies following Tregs infusion to access the localization of infused cells. Optimizing the manufacturing processes and culture media will support infused Tregs survival in future clinical trials. In addition, improving our understanding on the patient's omics profile with new technology will also allow us to put the personalized Tregs immunotherapy into effect.

In a word, although challenges still remain, the prospect of Tregs immunotherapy is exciting if the cell therapy community can maintain the collaboration closely. The immunosuppression-free period for patients with autoimmune disease and transplantation is in front of us.

Author details

Yang Liu[1,2*], Tiezheng Hou[1,2*] and Huiqin Hao[1,2*]

1 College of Basic Medical Sciences, Shanxi University of Chinese Medicine, Jinzhong, P.R. China

2 Basic Laboratory of Integrated Traditional Chinese and Western Medicine, Shanxi University of Chinese Medicine, Jinzhong, P.R. China

*Address all correspondence to: liuyang1980@sxtcm.edu.cn, th@sxtcm.edu.cn and hhq@sxtcm.edu.cn

References

[1] Gershon RK, Kondo K. Cell interactions in the induction of tolerance: The role of thymic lymphocytes. Immunology. 1970;**18**(5):723-737

[2] Zhang R, Miao J, Zhu P. Regulatory T cell heterogeneity and therapy in autoimmune diseases. Autoimmunity Reviews. 2021;**20**(5):102715

[3] Grover P, Goel PN, Greene MI. Regulatory T cells: Regulation of identity and function. Frontiers in Immunology. 2021;**12**:750542

[4] Fontenot JD, Gavin MA, Rudensky AY. Foxp3 programs the development and function of $CD4^{+}CD25^{+}$ regulatory T cells. Nature Immunology. 2003;**4**(4):330-336

[5] Hu W, Wang Z-M, Feng Y, Schizas M, Hoyos BE, van der Veeken J, et al. Regulatory T cells function in established systemic inflammation and reverse fatal autoimmunity. Nature Immunology. 2021;**22**(9):1163-1174

[6] Chatila TA, Blaeser F, Ho N, Lederman HM, Voulgaropoulos C, Helms C, et al. JM2, encoding a fork head-related protein, is mutated in X-linked autoimmunity-allergic disregulation syndrome. The Journal of Clinical Investigation. 2000;**106**(12):R75-R81

[7] Torgerson TR, Ochs HD. Immune dysregulation, polyendocrinopathy, enteropathy, X-linked: Forkhead box protein 3 mutations and lack of regulatory T cells. The Journal of Allergy and Clinical Immunology. 2007;**120**(4):744-750

[8] Verbsky JW, Chatila TA. Immune dysregulation, polyendocrinopathy, enteropathy, X-linked (IPEX) and IPEX-related disorders: An evolving web of heritable autoimmune diseases. Current Opinion in Pediatrics. 2013;**25**(6):708-714

[9] Attia P, Phan GQ, Maker AV, Robinson MR, Quezado MM, Yang JC, et al. Autoimmunity correlates with tumor regression in patients with metastatic melanoma treated with anti-cytotoxic T-lymphocyte antigen-4. Journal of Clinical Oncology. 2005;**23**(25):6043-6053

[10] Curiel TJ, Coukos G, Zou L, Alvarez X, Cheng P, Mottram P, et al. Specific recruitment of regulatory T cells in ovarian carcinoma fosters immune privilege and predicts reduced survival. Nature Medicine. 2004;**10**(9):942-949

[11] Antony PA, Piccirillo CA, Akpinarli A, Finkelstein SE, Speiss PJ, Surman DR, et al. $CD8^{+}$ T cell immunity against a tumor/self-antigen is augmented by $CD4^{+}$ T helper cells and hindered by naturally occurring T regulatory cells. Journal of Immunology. 2005;**174**(5):2591-2601

[12] Ménard C, Ghiringhelli F, Roux S, Chaput N, Mateus C, Grohmann U, et al. Ctla-4 blockade confers lymphocyte resistance to regulatory T-cells in advanced melanoma: Surrogate marker of efficacy of tremelimumab? Clinical Cancer Research. 2008;**14**(16):5242-5249

[13] Bennett CL, Christie J, Ramsdell F, Brunkow ME, Ferguson PJ, Whitesell L, et al. The immune dysregulation, polyendocrinopathy, enteropathy, X-linked syndrome (IPEX) is caused by mutations of FOXP3. Nature Genetics. 2001;**27**(1):20-21

[14] Brunkow ME, Jeffery EW, Hjerrild KA, Paeper B, Clark LB, Yasayko SA, et al.

Disruption of a new forkhead/winged-helix protein, scurfin, results in the fatal lymphoproliferative disorder of the scurfy mouse. Nature Genetics. 2001;**27**(1):68-73

[15] Chinen T, Kannan AK, Levine AG, Fan X, Klein U, Zheng Y, et al. An essential role for the IL-2 receptor in T cell function. Nature Immunology. 2016;**17**(11):1322-1333

[16] Sakaguchi S, Yamaguchi T, Nomura T, Ono M. Regulatory T cells and immune tolerance. Cell. 2008;**133**(5):775-787

[17] Read S, Malmström V, Powrie F. Cytotoxic T lymphocyte-associated antigen 4 plays an essential role in the function of CD25(+)CD4(+) regulatory cells that control intestinal inflammation. The Journal of Experimental Medicine. 2000;**192**(2):295-302

[18] Field EH, Kulhankova K, Nasr ME. Natural Tregs, CD4+CD25+ inhibitory hybridomas, and their cell contact dependent suppression. Immunologic Research. 2007;**39**(1-3):62-78

[19] Pandiyan P, Zheng L, Ishihara S, Reed J, Lenardo MJ. CD4+CD25+Foxp3+ regulatory T cells induce cytokine deprivation-mediated apoptosis of effector CD4+ T cells. Nature Immunology. 2007;**8**(12):1353-1362

[20] Gorelik L, Constant S, Flavell RA. Mechanism of transforming growth factor beta-induced inhibition of T helper type 1 differentiation. The Journal of Experimental Medicine. 2002;**195**(11):1499-1505

[21] Collison LW, Workman CJ, Kuo TT, Boyd K, Wang Y, Vignali KM, et al. The inhibitory cytokine IL-35 contributes to regulatory T-cell function. Nature. 2007;**450**(7169):566-569

[22] Wohler J, Bullard D, Schoeb T, Barnum S. LFA-1 is critical for regulatory T cell homeostasis and function. Molecular Immunology. 2009;**46**(11-12):2424-2428

[23] Holmgaard RB, Zamarin D, Munn DH, Wolchok JD, Allison JP. Indoleamine 2,3-dioxygenase is a critical resistance mechanism in antitumor T cell immunotherapy targeting CTLA-4. The Journal of Experimental Medicine. 2013;**210**(7):1389-1402

[24] Hou TZ, Qureshi OS, Wang CJ, Baker J, Young SP, Walker LSK, et al. A transendocytosis model of CTLA-4 function predicts its suppressive behavior on regulatory T cells. Journal of Immunology. 2015;**194**(5):2148-2159

[25] Zóka A, Barna G, Somogyi A, Műzes G, Oláh Á, Al-Aissa Z, et al. Extension of the $CD4^{+}Foxp3^{+}CD25(-/$ low) regulatory T-cell subpopulation in type 1 diabetes mellitus. Autoimmunity. 2015;**48**(5):289-297

[26] Schneider A, Rieck M, Sanda S, Pihoker C, Greenbaum C, Buckner JH. The effector T cells of diabetic subjects are resistant to regulation via CD4+ FOXP3+ regulatory T cells. Journal of Immunology. 2008;**181**(10):7350-7355

[27] Brusko T, Wasserfall C, McGrail K, Schatz R, Viener HL, Schatz D, et al. No alterations in the frequency of FOXP3+ regulatory T-cells in type 1 diabetes. Diabetes. 2007;**56**(3):604-612

[28] Sgouroudis E, Albanese A, Piccirillo CA. Impact of protective IL-2 allelic variants on CD4+ Foxp3+ regulatory T cell function in situ and resistance to autoimmune diabetes in NOD mice. Journal of Immunology. 2008;**181**(9):6283-6292

[29] Long SA, Cerosaletti K, Wan JY, Ho JC, Tatum M, Wei S, et al. An autoimmune-associated variant in

PTPN2 reveals an impairment of IL-2R signaling in CD4(+) T cells. Genes and Immunity. 2011;**12**(2):116-125

[30] Serr I, Fürst RW, Achenbach P, Scherm MG, Gökmen F, Haupt F, et al. Type 1 diabetes vaccine candidates promote human Foxp3(+)Treg induction in humanized mice. Nature Communications. 2016;7:10991

[31] Cabello-Kindelan C, Mackey S, Sands A, Rodriguez J, Vazquez C, Pugliese A, et al. Immunomodulation followed by antigen-specific Treg infusion controls islet autoimmunity. Diabetes. 2020;**69**(2):215-227

[32] Walter GJ, Evans HG, Menon B, Gullick NJ, Kirkham BW, Cope AP, et al. Interaction with activated monocytes enhances cytokine expression and suppressive activity of human CD4+CD45ro+CD25+CD127(low) regulatory T cells. Arthritis and Rheumatism. 2013;**65**(3):627-638

[33] Sun H, Gao W, Pan W, Zhang Q, Wang G, Feng D, et al. Tim3 Foxp3 Treg cells are potent inhibitors of effector T cells and are suppressed in rheumatoid arthritis. Inflammation. 2017;**40**(4):1342-1350

[34] Han GM, O'Neil-Andersen NJ, Zurier RB, Lawrence DA. CD4+CD25high T cell numbers are enriched in the peripheral blood of patients with rheumatoid arthritis. Cellular Immunology. 2008;**253**(1-2):92-101

[35] Möttönen M, Heikkinen J, Mustonen L, Isomäki P, Luukkainen R, Lassila O. CD4+ CD25+ T cells with the phenotypic and functional characteristics of regulatory T cells are enriched in the synovial fluid of patients with rheumatoid arthritis. Clinical and Experimental Immunology. 2005;**140**(2):360-367

[36] Lina C, Conghua W, Nan L, Ping Z. Combined treatment of etanercept and MTX reverses Th1/Th2, Th17/Treg imbalance in patients with rheumatoid arthritis. Journal of Clinical Immunology. 2011;**31**(4):596-605

[37] Wang J, Ioan-Facsinay A, van der Voort EIH, Huizinga TWJ, Toes REM. Transient expression of FOXP3 in human activated nonregulatory CD4+ T cells. European Journal of Immunology. 2007;**37**(1):129-138

[38] Schmetterer KG, Neunkirchner A, Pickl WF. Naturally occurring regulatory T cells: Markers, mechanisms, and manipulation. The FASEB Journal. 2012;**26**(6):2253-2276

[39] Machura E, Mazur B, Pieniazek W, Karczewska K. Expression of naive/memory (CD45RA/CD45RO) markers by peripheral blood CD4+ and CD8 + T cells in children with asthma. Archivum Immunologiae et Therapiae Experimentalis (Warsz). 2008;**56**(1):55-62

[40] Morita T, Shima Y, Wing JB, Sakaguchi S, Ogata A, Kumanogoh A. The proportion of regulatory T cells in patients with rheumatoid arthritis: A meta-analysis. PLoS One. 2016;**11**(9): e0162306

[41] Cao D, van Vollenhoven R, Klareskog L, Trollmo C, Malmström V. CD25brightCD4+ regulatory T cells are enriched in inflamed joints of patients with chronic rheumatic disease. Arthritis Research & Therapy. 2004;**6**(4):R335-RR46

[42] Li S, Yin H, Zhang K, Wang T, Yang Y, Liu X, et al. Effector T helper cell populations are elevated in the bone marrow of rheumatoid arthritis patients and correlate with disease severity. Scientific Reports. 2017;7(1):4776

[43] Rossetti M, Spreafico R, Consolaro A, Leong JY, Chua C, Massa M, et al. TCR repertoire sequencing identifies synovial Treg cell clonotypes in the bloodstream during active inflammation in human arthritis. Annals of the Rheumatic Diseases. 2017;**76**(2):435-441

[44] de Kleer IM, Wedderburn LR, Taams LS, Patel A, Varsani H, Klein M, et al. CD4+CD25bright regulatory T cells actively regulate inflammation in the joints of patients with the remitting form of juvenile idiopathic arthritis. Journal of Immunology. 2004;**172**(10):6435-6443

[45] Komatsu N, Okamoto K, Sawa S, Nakashima T, Oh-hora M, Kodama T, et al. Pathogenic conversion of Foxp3+ T cells into TH17 cells in autoimmune arthritis. Nature Medicine. 2014;**20**(1): 62-68

[46] Mieli-Vergani G, Vergani D, Czaja AJ, Manns MP, Krawitt EL, Vierling JM, et al. Autoimmune hepatitis. Nature Reviews. Disease Primers. 2018;**4**:18017

[47] Webb GJ, Hirschfield GM, Krawitt EL, Gershwin ME. Cellular and molecular mechanisms of autoimmune hepatitis. Annual Review of Pathology. 2018;**13**:247-292

[48] Beringer A, Miossec P. IL-17 and IL-17-producing cells and liver diseases, with focus on autoimmune liver diseases. Autoimmunity Reviews. 2018;**17**(12):1176-1185

[49] Gregorio GV, Portmann B, Karani J, Harrison P, Donaldson PT, Vergani D, et al. Autoimmune hepatitis/sclerosing cholangitis overlap syndrome in childhood: A 16-year prospective study. Hepatology. 2001;**33**(3):544-553

[50] Longhi MS, Ma Y, Bogdanos DP, Cheeseman P, Mieli-Vergani G, Vergani D. Impairment of CD4(+) CD25(+) regulatory T-cells in autoimmune liver disease. Journal of Hepatology. 2004;**41**(1):31-37

[51] Longhi MS, Ma Y, Mitry RR, Bogdanos DP, Heneghan M, Cheeseman P, et al. Effect of CD4+ CD25+ regulatory T-cells on CD8 T-cell function in patients with autoimmune hepatitis. Journal of Autoimmunity. 2005;**25**(1):63-71

[52] Longhi MS, Mitry RR, Samyn M, Scalori A, Hussain MJ, Quaglia A, et al. Vigorous activation of monocytes in juvenile autoimmune liver disease escapes the control of regulatory T-cells. Hepatology. 2009;**50**(1):130-142

[53] Grant CR, Liberal R, Holder BS, Cardone J, Ma Y, Robson SC, et al. Dysfunctional CD39(POS) regulatory T cells and aberrant control of T-helper type 17 cells in autoimmune hepatitis. Hepatology. 2014;**59**(3):1007-1015

[54] Vuerich M, Harshe R, Frank LA, Mukherjee S, Gromova B, Csizmadia E, et al. Altered aryl-hydrocarbon-receptor signalling affects regulatory and effector cell immunity in autoimmune hepatitis. Journal of Hepatology. 2021;**74**(1):48-57

[55] Liberal R, Grant CR, Holder BS, Ma Y, Mieli-Vergani G, Vergani D, et al. The impaired immune regulation of autoimmune hepatitis is linked to a defective galectin-9/tim-3 pathway. Hepatology. 2012;**56**(2):677-686

[56] Diestelhorst J, Junge N, Schlue J, Falk CS, Manns MP, Baumann U, et al. Pediatric autoimmune hepatitis shows a disproportionate decline of regulatory T cells in the liver and of IL-2 in the blood of patients undergoing therapy. PLoS One. 2017;**12**(7):e0181107

[57] Ananthakrishnan AN. Epidemiology and risk factors for IBD. Nature Reviews.

Gastroenterology & Hepatology. 2015;**12**(4):205-217

[58] Iacomino G, Rotondi Aufiero V, Iannaccone N, Melina R, Giardullo N, De Chiara G, et al. IBD: Role of intestinal compartments in the mucosal immune response. Immunobiology. 2020;**225**(1):151849

[59] Ueno A, Jeffery L, Kobayashi T, Hibi T, Ghosh S, Jijon H. Th17 plasticity and its relevance to inflammatory bowel disease. Journal of Autoimmunity. 2018;**87**:38-49

[60] Veldhoen M, Hocking RJ, Atkins CJ, Locksley RM, Stockinger B. TGFbeta in the context of an inflammatory cytokine milieu supports de novo differentiation of IL-17-producing T cells. Immunity. 2006;**24**(2):179-189

[61] Acharya S, Timilshina M, Jiang L, Neupane S, Choi D-Y, Park SW, et al. Amelioration of experimental autoimmune encephalomyelitis and DSS induced colitis by NTG-A-009 through the inhibition of Th1 and Th17 cells differentiation. Scientific Reports. 2018;**8**(1):7799

[62] Elmore JP, McGee MC, Nidetz NF, Anannya O, Huang W, August A. Tuning T helper cell differentiation by ITK. Biochemical Society Transactions. 2020;**48**(1):179-185

[63] Zhang Z, Zhong W, Hinrichs D, Wu X, Weinberg A, Hall M, et al. Activation of OX40 augments Th17 cytokine expression and antigen-specific uveitis. The American Journal of Pathology. 2010;**177**(6):2912-2920

[64] Fan Z, Spencer JA, Lu Y, Pitsillides CM, Singh G, Kim P, et al. In vivo tracking of 'color-coded' effector, natural and induced regulatory T cells in the allograft response. Nature Medicine. 2010;**16**(6):718-722

[65] Zhang N, Schröppel B, Lal G, Jakubzick C, Mao X, Chen D, et al. Regulatory T cells sequentially migrate from inflamed tissues to draining lymph nodes to suppress the alloimmune response. Immunity. 2009;**30**(3):458-469

[66] Game DS, Lechler RI. Pathways of allorecognition: Implications for transplantation tolerance. Transplant Immunology. 2002;**10**(2-3):101-108

[67] Wood KJ, Sakaguchi S. Regulatory T cells in transplantation tolerance. Nature Reviews. Immunology. 2003;**3**(3):199-210

[68] Beilhack A, Schulz S, Baker J, Beilhack GF, Wieland CB, Herman EI, et al. In vivo analyses of early events in acute graft-versus-host disease reveal sequential infiltration of T-cell subsets. Blood. 2005;**106**(3):1113-1122

[69] Taylor PA, Noelle RJ, Blazar BR. CD4(+)CD25(+) immune regulatory cells are required for induction of tolerance to alloantigen via costimulatory blockade. The Journal of Experimental Medicine. 2001;**193**(11):1311-1318

[70] Trenado A, Charlotte F, Fisson S, Yagello M, Klatzmann D, Salomon BL, et al. Recipient-type specific CD4+CD25+ regulatory T cells favor immune reconstitution and control graft-versus-host disease while maintaining graft-versus-leukemia. The Journal of Clinical Investigation. 2003;**112**(11):1688-1696

[71] Edinger M, Hoffmann P, Ermann J, Drago K, Fathman CG, Strober S, et al. CD4+CD25+ regulatory T cells preserve graft-versus-tumor activity while inhibiting graft-versus-host disease after bone marrow transplantation. Nature Medicine. 2003;**9**(9):1144-1150

[72] Nguyen VH, Shashidhar S, Chang DS, Ho L, Kambham N, Bachmann M, et al.

The impact of regulatory T cells on T-cell immunity following hematopoietic cell transplantation. Blood. 2008;**111**(2): 945-953

[73] Negrin RS, Contag CH. In vivo imaging using bioluminescence: A tool for probing graft-versus-host disease. Nature Reviews. Immunology. 2006;**6**(6):484-490

[74] Lin C-H, Hünig T. Efficient expansion of regulatory T cells in vitro and in vivo with a CD28 superagonist. European Journal of Immunology. 2003;**33**(3):626-638

[75] Hoffmann P, Eder R, Kunz-Schughart LA, Andreesen R, Edinger M. Large-scale in vitro expansion of polyclonal human CD4(+) CD25high regulatory T cells. Blood. 2004;**104**(3):895-903

[76] Sakaguchi S, Sakaguchi N, Asano M, Itoh M, Toda M. Immunologic self-tolerance maintained by activated T cells expressing IL-2 receptor alpha-chains (CD25). Breakdown of a single mechanism of self-tolerance causes various autoimmune diseases. Journal of Immunology. 1995;**155**(3):1151-1164

[77] Taylor PA, Lees CJ, Blazar BR. The infusion of ex vivo activated and expanded CD4(+)CD25(+) immune regulatory cells inhibits graft-versus-host disease lethality. Blood. 2002;**99**(10):3493-3499

[78] Xia G, He J, Leventhal JR. Ex vivo-expanded natural CD4+CD25+ regulatory T cells synergize with host T-cell depletion to promote long-term survival of allografts. American Journal of Transplantation. 2008;**8**(2):298-306

[79] Canavan JB, Scottà C, Vossenkämper A, Goldberg R, Elder MJ, Shoval I, et al. Developing in vitro expanded CD45RA+ regulatory T cells as an adoptive cell therapy for Crohn's disease. Gut. 2016;**65**(4):584-594

[80] Bluestone JA, Buckner JH, Fitch M, Gitelman SE, Gupta S, Hellerstein MK, et al. Type 1 diabetes immunotherapy using polyclonal regulatory T cells. Science Translational Medicine. 2015;**7**(315):315ra189

[81] Mathew JM, H-Voss J, LeFever A, Konieczna I, Stratton C, He J, et al. A phase I clinical trial with ex vivo expanded recipient regulatory T cells in living donor kidney transplants. Scientific Reports. 2018;**8**(1):7428

[82] Trzonkowski P, Bieniaszewska M, Juścińska J, Dobyszuk A, Krzystyniak A, Marek N, et al. First-in-man clinical results of the treatment of patients with graft versus host disease with human ex vivo expanded CD4+CD25+CD127− T regulatory cells. Clinical Immunology. 2009;**133**(1):22-26

[83] Marek-Trzonkowska N, Myśliwiec M, Dobyszuk A, Grabowska M, Derkowska I, Juścińska J, et al. Therapy of type 1 diabetes with CD4(+) CD25(high)CD127-regulatory T cells prolongs survival of pancreatic islets—Results of one year follow-up. Clinical Immunology. 2014;**153**(1):23-30

[84] Esensten JH, Muller YD, Bluestone JA, Tang Q. Regulatory T-cell therapy for autoimmune and autoinflammatory diseases: The next frontier. The Journal of Allergy and Clinical Immunology. 2018;**142**(6):1710-1718

[85] Rosenzwajg M, Lorenzon R, Cacoub P, Pham HP, Pitoiset F, El Soufi K, et al. Immunological and clinical effects of low-dose interleukin-2 across 11 autoimmune diseases in a single, open clinical trial. Annals of the Rheumatic Diseases. 2019;**78**(2):209-217

[86] Lim TY, Martinez-Llordella M, Kodela E, Gray E, Heneghan MA, Sanchez-Fueyo A. Low-dose Interleukin-2 for refractory autoimmune hepatitis. Hepatology. 2018;**68**(4):1649-1652

[87] McDermott DF. The application of high-dose interleukin-2 for metastatic renal cell carcinoma. Medical Oncology. 2009;**26**(Suppl. 1):13-17

[88] Zanin-Zhorov A, Cahalon L, Tal G, Margalit R, Lider O, Cohen IR. Heat shock protein 60 enhances CD4+CD25+ regulatory T cell function via innate TLR2 signaling. The Journal of Clinical Investigation. 2018;**128**(6):2651

[89] Massa M, Passalia M, Manzoni SM, Campanelli R, Ciardelli L, Yung GP, et al. Differential recognition of heat-shock protein dnaJ-derived epitopes by effector and Treg cells leads to modulation of inflammation in juvenile idiopathic arthritis. Arthritis and Rheumatism. 2007;**56**(5):1648-1657

[90] Sutmuller R, Garritsen A, Adema GJ. Regulatory T cells and toll-like receptors: Regulating the regulators. Annals of the Rheumatic Diseases. 2007;**66** (Suppl. 3):iii91-iiii5

[91] Aalberse JA, Kapitein B, de Roock S, Klein MR, de Jager W, van der Zee R, et al. Cord blood CD4+ T cells respond to self heat shock protein 60 (HSP60). PLoS One. 2011;**6**(9):e24119

[92] Wachstein J, Tischer S, Figueiredo C, Limbourg A, Falk C, Immenschuh S, et al. HSP70 enhances immunosuppressive function of CD4(+)CD25(+)FoxP3(+) T regulatory cells and cytotoxicity in CD4(+)CD25(−) T cells. PLoS One. 2012;7(12):e51747

[93] Prakken BJ, Wendling U, van der Zee R, Rutten VP, Kuis W, van Eden W. Induction of IL-10 and inhibition of experimental arthritis are specific features of microbial heat shock proteins that are absent for other evolutionarily conserved immunodominant proteins. Journal of Immunology. 2001;**167**(8):4147-4153

[94] van Eden W, Hauet-Broere F, Berlo S, Paul L, van der Zee R, de Kleer I, et al. Stress proteins as inducers and targets of regulatory T cells in arthritis. International Reviews of Immunology. 2005;**24**(3-4):181-197

[95] van Herwijnen MJC, Wieten L, van der Zee R, van Kooten PJ, Wagenaar-Hilbers JP, Hoek A, et al. Regulatory T cells that recognize a ubiquitous stress-inducible self-antigen are long-lived suppressors of autoimmune arthritis. Proceedings of the National Academy of Sciences of the United States of America. 2012;**109**(35):14134-14139

[96] de Zoeten EF, Wang L, Butler K, Beier UH, Akimova T, Sai H, et al. Histone deacetylase 6 and heat shock protein 90 control the functions of Foxp3(+) T-regulatory cells. Molecular and Cellular Biology. 2011;**31**(10):2066-2078

[97] Zhang Y, Wu BX, Metelli A, Thaxton JE, Hong F, Rachidi S, et al. GP96 is a GARP chaperone and controls regulatory T cell functions. The Journal of Clinical Investigation. 2015;**125**(2):859-869

[98] Xu Y, Liu E, Xie X, Wang J, Zheng H, Ju Y, et al. Induction of Foxp3 and activation of Tregs by HSP gp96 for treatment of autoimmune diseases. iScience. 2021;**24**(12):103445

[99] Hori S, Nomura T, Sakaguchi S. Control of regulatory T cell development by the transcription factor Foxp3. Science. 2003;**299**(5609):1057-1061

[100] Passerini L, Rossi Mel E, Sartirana C, Fousteri G, Bondanza A, Naldini L, et al. CD4+ T cells from IPEX patients convert into functional and stable regulatory T cells by FOXP3 gene transfer. Science Translational Medicine. 2013;**5**(215):215ra174

[101] Loser K, Hansen W, Apelt J, Balkow S, Buer J, Beissert S. In vitro-generated regulatory T cells induced by Foxp3-retrovirus infection control murine contact allergy and systemic autoimmunity. Gene Therapy. 2005;**12**(17):1294-1304

[102] Andersen KG, Butcher T, Betz AG. Specific immunosuppression with inducible Foxp3-transduced polyclonal T cells. PLoS Biology. 2008;**6**(11):e276

[103] Honaker Y, Sommer K, Dahl N, Xiang Y, Rawlings DJ. Enabling gene-edited, regulatory-like, T cells (edTreg) for treatment of IPEX and other autoimmune disorders. Blood. 2019;**134**(Supplement_1):2071

[104] Liu X, Xu X, Lin X, Tian Y, Ji B, Xia S, et al. PTD-hFOXP3 protein acts as an immune regulator to convert human CD4(+)CD25(−) T cells to regulatory T-like cells. Journal of Cellular Biochemistry. 2012;**113**(12):3797-3809

[105] Wing JB, Tanaka A, Sakaguchi S. Human FOXP3 regulatory T cell heterogeneity and function in autoimmunity and cancer. Immunity. 2019;**50**(2):302-316

[106] Gottschalk RA, Corse E, Allison JP. TCR ligand density and affinity determine peripheral induction of Foxp3 in vivo. The Journal of Experimental Medicine. 2010;**207**(8):1701-1711

[107] Daniel C, Weigmann B, Bronson R, von Boehmer H. Prevention of type 1 diabetes in mice by tolerogenic vaccination with a strong agonist insulin mimetope. The Journal of Experimental Medicine. 2011;**208**(7):1501-1510

[108] Tang Q, Henriksen KJ, Bi M, Finger EB, Szot G, Ye J, et al. In vitro-expanded antigen-specific regulatory T cells suppress autoimmune diabetes. The Journal of Experimental Medicine. 2004;**199**(11):1455-1465

[109] Sagoo P, Ali N, Garg G, Nestle FO, Lechler RI, Lombardi G. Human regulatory T cells with alloantigen specificity are more potent inhibitors of alloimmune skin graft damage than polyclonal regulatory T cells. Science Translational Medicine. 2011;**3**(83):83ra42

[110] Green EA, Choi Y, Flavell RA. Pancreatic lymph node-derived CD4(+) CD25(+) Treg cells: Highly potent regulators of diabetes that require TRANCE-RANK signals. Immunity. 2002;**16**(2):183-191

[111] Blat D, Zigmond E, Alteber Z, Waks T, Eshhar Z. Suppression of murine colitis and its associated cancer by carcinoembryonic antigen-specific regulatory T cells. Molecular Therapy. 2014;**22**(5):1018-1028

[112] Jethwa H, Adami AA, Maher J. Use of gene-modified regulatory T-cells to control autoimmune and alloimmune pathology: Is now the right time? Clinical Immunology. 2014;**150**(1):51-63

[113] Wright GP, Notley CA, Xue S-A, Bendle GM, Holler A, Schumacher TN, et al. Adoptive therapy with redirected primary regulatory T cells results in antigen-specific suppression of arthritis. Proceedings of the National Academy of Sciences of the United States of America. 2009;**106**(45):19078-19083

[114] Tsang JY-S, Tanriver Y, Jiang S, Xue S-A, Ratnasothy K, Chen D, et al.

Conferring indirect allospecificity on CD4+CD25+ Tregs by TCR gene transfer favors transplantation tolerance in mice. The Journal of Clinical Investigation. 2008;**118**(11):3619-3628

[115] Wang GC, Dash P, McCullers JA, Doherty PC, Thomas PG. T cell receptor αβ diversity inversely correlates with pathogen-specific antibody levels in human cytomegalovirus infection. Science Translational Medicine. 2012;**4**(128):128ra42

[116] Spence A, Purtha W, Tam J, Dong S, Kim Y, Ju C-H, et al. Revealing the specificity of regulatory T cells in murine autoimmune diabetes. Proceedings of the National Academy of Sciences of the United States of America. 2018;**115**(20):5265-5270

[117] Boardman D, Maher J, Lechler R, Smyth L, Lombardi G. Antigen-specificity using chimeric antigen receptors: The future of regulatory T-cell therapy? Biochemical Society Transactions. 2016;**44**(2):342-348

[118] MacDonald KG, Hoeppli RE, Huang Q, Gillies J, Luciani DS, Orban PC, et al. Alloantigen-specific regulatory T cells generated with a chimeric antigen receptor. The Journal of Clinical Investigation. 2016;**126**(4):1413-1424

[119] Elinav E, Waks T, Eshhar Z. Redirection of regulatory T cells with predetermined specificity for the treatment of experimental colitis in mice. Gastroenterology. 2008;**134**(7):2014-2024

[120] Elinav E, Adam N, Waks T, Eshhar Z. Amelioration of colitis by genetically engineered murine regulatory T cells redirected by antigen-specific chimeric receptor. Gastroenterology. 2009;**136**(5):1721-1731

[121] De Paula PA, Schmidt A, Zhang AH, Maldonado T, Konigs C, Scott DW. Engineered regulatory T cells expressing myelin-specific chimeric antigen receptors suppress EAE progression. Cellular Immunology. 2020;**358**:104222

[122] Tarbell KV, Yamazaki S, Olson K, Toy P, Steinman RM. CD25+ CD4+ T cells, expanded with dendritic cells presenting a single autoantigenic peptide, suppress autoimmune diabetes. The Journal of Experimental Medicine. 2004;**199**(11):1467-1477

[123] Fujio K, Okamoto A, Araki Y, Shoda H, Tahara H, Tsuno NH, et al. Gene therapy of arthritis with TCR isolated from the inflamed paw. Journal of Immunology. 2006;**177**(11):8140-8147

[124] Honaker Y, Hubbard N, Xiang Y, Fisher L, Hagin D, Sommer K, et al. Gene editing to induce FOXP3 expression in human CD4 T cells leads to a stable regulatory phenotype and function. Science Translational Medicine. 2020;**12**(546):eaay6422

[125] Boardman DA, Philippeos C, Fruhwirth GO, Ibrahim MAA, Hannen RF, Cooper D, et al. Expression of a chimeric antigen receptor specific for donor HLA class I enhances the potency of human regulatory T cells in preventing human skin transplant rejection. American Journal of Transplantation. 2017;**17**(4):931-943

[126] Huter EN, Stummvoll GH, DiPaolo RJ, Glass DD, Shevach EM. Cutting edge: Antigen-specific TGF beta-induced regulatory T cells suppress Th17-mediated autoimmune disease. Journal of Immunology. 2008;**181**(12):8209-8213

[127] Semple K, Yu Y, Wang D, Anasetti C, Yu X-Z. Efficient and selective prevention of GVHD by antigen-specific induced Tregs via linked-suppression in mice. Biology of Blood and Marrow Transplantation. 2011;**17**(3):309-318

Chapter 3

The Role of Immune Checkpoints in Cancer Progression

Rahmad Aji Prasetya and Devyani Diah Wulansari

Abstract

Immune checkpoint proteins are like two-faced swords that first act as gatekeepers of the immune system to protect the host from tissue damage. In contrast, these proteins can corroborate cancer progression by inhibiting tumor-specific immune responses. Here, we summarized the regulation and signaling cascade of immune checkpoints molecules (PD-1/PD-L1, CTLA-4, TIM3, TIGIT, LAG3, and BTLA), including their role in providing co-inhibitory signals for regulating T-cell response. The involvement of immune checkpoint molecules to drive cancer growth is elaborated with explanations about various anticancer strategies, such as (1) the overexpression of immune checkpoints in cancer cells, immune cells, or the surrounding environment leading to incapabilities of the tumor-specific immune response, (2) immune checkpoints interference to metabolic pathways then deplete nutrients needed by immune cells, (3) the interaction between immune checkpoints and regulatory T cells. Lastly, future challenges of immune checkpoint inhibitors are discussed briefly to get insight into their applicability in the clinical setting.

Keywords: immune checkpoint proteins, cancer development, anti-tumor, metabolic reprogramming, regulatory T cells

1. Introduction

Cancer or tumor cells express neoantigens that the immune system can identify from healthy neighboring cells due to genetic mutations. These changes typically result in a tumor-reactive T cell response, most notably $CD8^+$ T cells. However, this mechanism is frequently ineffective at eradicating cancer cells [1]. One cause for this failure is the suppression of invading T cells by a wide range of immunosuppressive mechanisms found in the tumor microenvironment (TME), such as regulatory T cells (Tregs) or immunosuppressive cytokines [2, 3].

Furthermore, binding of the T cell receptor (TCR) to the antigenic peptide bound to the major histocompatibility complex (MHC) of the antigen-presenting cell (APC) is not adequate to yield an immune response, particularly to eradicate cancer cells. Thus, the additional stimulatory co-signal produced by co-receptors is required. These co-receptors play an essential role in modulating T cell responsiveness and balancing co-stimulatory and inhibitory (i.e., immune checkpoint) signals [4]. Extended TCR signals generated from T cells exposure to their cognate antigen result in enhanced and persistent expression of inhibitory co-receptors like cytotoxic T-lymphocyte-associated antigen 4 (CTLA-4), programmed death protein 1 (PD-1), or many other immune checkpoints. At this moment, T cells enter a state of dysfunction or exhaustion, allowing cancer cells to grow unchecked [5, 6]. Therefore, blockage to these immune checkpoints can reinvigorate the anti-tumor function of immune cells. This chapter aimed to elaborate on the involvement of immune checkpoints in cancer development. It includes the explanation of the normal trafficking and inhibitory signaling of each checkpoint, followed by discussions about how immune checkpoint contributes to cancer growth.

2. Regulation and signaling of immune checkpoints

Immune checkpoints serve as the immune system's gatekeepers and are required for sustaining self-tolerance, thus protecting the host from tissue damage. These immunological checkpoint molecules have modulated T cell responses to self-proteins, persistent infections, and tumor antigens. A few of them, including but are not limited to PD-1, CTLA-4, Lymphocyte activation gene 3 (LAG3; or known as cluster of differentiation 223 [CD223]), T cell immunoglobulin and mucin-domain containing-3 (TIM-3), T cell immunoreceptor with immunoglobulin and ITIM—immunoreceptor tyrosine-based inhibitory motif—domain (TIGIT), and B and T lymphocyte attenuator (BTLA; or known as CD272), have been discovered and investigated as targets in cancer immunotherapy. In general, immune checkpoints are membrane proteins expressed in the endoplasmic reticulum (ER) and subsequently transported to the cell surface to perform their inhibitory roles, which requires the protein sorting system to transport them sequentially through the Golgi apparatus secretory vesicles. Glycosylation acts as quality control during surface delivery, ensuring that only mature and functional immunological checkpoints reach the cell surface. Immune checkpoints are internalized and recycled when they reach the cell surface, providing a quick regulatory pathway to control their surface levels. Immune checkpoints can be ubiquitinated and sorted to the proteasome or lysosome for destruction, another critical method for controlling protein levels. The surface level of immunological checkpoints is determined by several biological mechanisms, which affect cell signaling [7]. This section elaborates on the normal regulations and signaling of each immune checkpoints molecules before discussing its involvement in cancer development.

2.1 PD-1/PD-L1 regulations and signaling

PD-1 trafficking in the membrane is regulated by the core fucosyltransferase 8 (fut8) in ER. Upon Tcell activation, PD-1 is internalized, then ubiquitinated by F-box protein 38 (FBXO38) for proteasome degradation or recycled back to the surface with the help of thymocyte selection-associated high mobility group box protein (Tox), thus prolonged PD-1 activity. Additionally, Tox expression induces Tcell depletion in hepatocellular cancer [7, 8]. Besides, another extension of PD-1 activity is caused by FBXO38's low transcriptional level in the TME. Hence, the FBXO38-mediated PD-1 degradation pathway is defective. TCR signaling was the source of FBXO38 downregulation in the absence of concurrent CD28-CD80/86 signaling. CD28-CD80/86 binding provides critical signals for T cell activation in the presence of TCR stimulation. Persistent tumor antigen binding and low CD80/86 expression on cancerous cells might explain the lower FBXO38 expression in tumor-infiltrating lymphocytes (TILs) [7, 9].

Similar to PD-1, its first functionally identified ligand of PD-L1 (also known as B7 homolog 1 [B7-H1] or CD274) is constantly internalized, recycled, or degraded.

Regulation of PD-L1 recycling is managed by CKLF-like MARVEL transmembrane domain containing 6 (CMTM6). Meanwhile, ubiquitination and degradation are regulated by multiple proteins such as cyclin D–CDK4 and the cullin 3–SPOP [10], β-TrCP [11], COP9 signalosome 5 (CSN5) [12], Huntingtin-interacting protein 1-related (HIP1R) [13], and others. Each protein is a drugable target to inhibit PD-L1 accumulation, thereby increasing T cell-mediated cytotoxicity.

Regarding the inhibitory signals following the binding of PD1 to PD-L1 or other ligands, it blocks kinases that play a role in activating T cells through the phosphatase SHP2. Besides, since PD1 inhibition blocks the TCR 'stop signal', this pathway can alter the length of T cell–APC or T cell–target cell interaction [14]. In detail, PD-1 is phosphorylated through immune receptor tyrosine-based switch motif (ITSM) and ITIM. Then, PD-1 binds the Src homology 2 (SH2) domains of SH2-containing phosphatase 2 (SHP2) or SHP1, which initiate its inhibitory effect by suppressing both TCR and CD28 co-stimulatory signaling [7, 15–17]. Moreover, PD-1 signaling also reduces cytokine production (interleukin [IL]-2, interferon [IFN]- α, tumor necrosis factor [TNF]-α), cell cycle progression, and pro-survival Bcl-xL gene expression by interfering with early TCR/CD28 signaling. PD-1/PD-L1 interaction is associated with IL-2-dependent positive feedback and transcription factors involved in effector functions such as GATA-3, T-bet, and Eomes. As signal transduction can only occur during TCR-dependent signaling, PD-1 activity is thus only relevant during simultaneous T cell activation. Mice without the receptor appear healthy at first. Still, they acquire autoimmune disorders such as lupus-like proliferative glomerulonephritis and arthritis, as well as enhanced inflammation after infections at a later age. In humans, genetic variations in the PD-1 region are more likely to suffer autoimmune disorders [18, 19].

2.2 CTLA-4 regulations and signaling

Unlike PD-1/PD-L1, which is constitutively expressed on the membrane, CTLA-4 is primarily stored inside the cytoplasm of resting naïve T lymphocytes. The T cell receptor-interacting molecule (TRIM)/LAX/Rab8 complex and phospholipase D (PLD)/ADP ribosylation factor-1 (ARF1)-dependent exocytosis are required for CTLA-4 trafficking from trans Golgi network (TGN) to the cell surface [20]. Exocytosis of CTLA-4-containing vesicles causes upregulation of CTLA-4 on the cell surface due to stimulatory signals originating from TCR and CD28-B7 interaction. More robust TCR signaling causes more CTLA-4 to be translocated to the cell surface, and this process works in a graded feedback loop. CTLA-4 on the surface is rapidly internalized during normal physiologic conditions, resulting in relatively low expression. The clathrin-associated adaptor complex (AP-2) interaction to the unphosphorylated YVKM motif promotes rapid CTLA-4 internalization, which is then either destroyed in the lysosome or returned to the cell surface through LPS responsive beige-like anchor protein (LRBA). Besides, CTLA-4 in TGN may also be transported to the lysosome for destruction through AP-1 binding [7, 21, 22].

The intrinsic signaling of CTLA-4 that dampens T cell immune response has been widely contested with no agreement [23]. However, both CTLA-4 and CD28 interact with the identical ligands, CD80 (B7–1) and CD86 (B7–2). Because CTLA-4 has a 20-fold higher binding affinity than CD28, the intrinsic inhibitory signal rises once CTLA-4 outcompetes CD28, even if CTLA-4 is activated later [24, 25]. In addition to T cell response intrinsic inhibition, CTLA-4 is hypothesized to decrease extrinsic T cell signaling. For example, CTLA-4 suppresses CD80/86 expression on APCs via trans-endocytosis or by increasing tumor growth factor β (TGFβ), which in turn suppresses CD80/86 expression [26]. CTLA-4 is phosphorylated when it binds to its ligands, activating phosphoinositide 3-kinase (PI3K) pathways leading to dephosphorylation of the CD3 chain, decreasing the TCR's signaling potential. CTLA-4 also prevents T cells from proliferating by inhibiting IL-2 transcription. Additionally, CTLA-4 stimulates the production of indoleamine 2,3-dioxygenase (IDO) in dendritic cells via CD80/86 ligation, resulting in T cell suppression [27].

2.3 TIM-3 regulations and signaling

TIM-3 is expressed on both T cells and innate immune cells. Four ligands have been identified: carcinoembryonic antigen cell adhesion molecule 1 (Ceacam1), C-type lectin galectin9 (Galectin9), high-mobility group box 1 (HMGB1), and non-protein ligand phosphatidylserine (PtdSer). Ceacam1 is a transmembrane protein that interacts in *cis* and *trans* directions. Ceacam1-TIM-3 *cis* binding induces TIM3 surface expression in T cells while *trans* binding inhibits the effector T cell causing exhaustion and maintaining T cell tolerance [28]. TIM-3 binding to both Ceacam1 and galectin-9 results in the release of Bat3, a TIM-3 signaling pathway inhibitory regulator, from its binding location on the Tim-3 cytoplasmic tail [29]. The other ligands, HMGB1, primarily modulate innate immunity like dendritic cells (DC). In DCs, HMGB1 is required for nucleic acid trafficking into endosomal vesicles, a fundamental step in sensing tumor-derived stressors or pathogen-associated molecular patterns and initiating host defenses against malignancies or pathogen infections [30].

TIM-3 is more related to co-stimulatory proteins induced in activated T cells than to a dominant inhibitory protein like PD-1; thus, TIM-3 signaling remains a matter of debate. As checkpoint proteins, TIM-3 is a repressor of IFN-γ-secreting $CD4^+$ Th1 and $CD8^+$ T cells. These findings confirmed that inhibiting TIM-3 might correct the defective phenotype of T cells in vivo. In contrast, TIM-3 lacks a conventional ITIM or ITSM in its intracellular domain and lacks structural features that facilitate the recruitment of inhibitory phosphatases. Rather than that, both murine and human TIM-3 cytoplasmic tails include five conserved tyrosine residues, two of which, Y_{256} and Y_{263} in mice (Y_{265}, Y_{272} in humans), have been demonstrated to be crucial for coupling to downstream signaling pathways. Y_{256} and Y_{263} in TIM-3's C-terminal tail interact with Bat3 in the absence of ligand-mediated TIM-3 signaling. Bat3 binds the catalytically active form of Lck in this state, resulting in the formation of an intracellular molecular complex with TIM-3 that retains and maybe enhances T cell signaling while repressing TIM-3-mediated cell death and exhaustion [31, 32]. TIM-3 activation on exhausted effector T cells is closely attributed to PD-1 expression, confirming the functional relationship between TIM-3 and PD-1 throughout the development of T cell exhaustion [33]. Concomitant therapy of anti-TIM-3 and anti-PD-1 is significantly more successful in these models, resulting in more significant tumor regression than either TIM-3 or PD-1 inhibition alone. TIM-3 inhibition in the setting of adaptive resistance to PD-1 treatment may be a useful way to treat individuals who develop resistance to anti-PD-1 therapy. This therapy regimen may be particularly beneficial for malignancies with resistance and immune escape from PD-1 inhibition [34, 35].

2.4 TIGIT regulations and signaling

Like CTLA-4 and CD28, TIGIT and CD226 can interact with identical ligands, CD112 and CD155. TIGIT is a co-inhibitory receptor, while CD226 is a co-stimulatory receptor. Nevertheless, TIGIT possesses a higher affinity to its ligands than CD226; thus, TIGIT can inhibit co-stimulation signals by outcompeting CD226 ligands binding. TIGIT can bind directly to CD226 in *cis*, disrupting its homodimer formation and co-stimulatory activity [36, 37].

TIGIT's signaling is mostly studied in natural killer (NK) cells and activated CD4 and CD8 T cells. The cytoplasmic region of TIGIT comprises an ITIM motif and an immunoglobulin tail tyrosine (ITT)-like motif. Several studies demonstrate that tyrosine (Tyr_{225}) phosphorylation in either the ITIM or ITT-like motif is required for TIGIT's inhibitory action in human NK cells. According to Liu et al. (2013), the ITT-like motif recruits Src homology domain-containing inositol phosphatases (SHIP1) via cytosolic adaptor proteins Grb2. Recruited SHIP1 then suppresses phosphatidylinositol 3-kinase (PI3K) and mitogen-activated protein kinase (MAPK) signal to abolish NK cell function. Furthermore, TIGIT signaling can modulate the IFN-γ production of NK cells via the NF-κB pathway. In this context, β-arrestin 2, another TIGIT adaptor, is involved in phosphorylating TIGIT and then inhibits TNF receptor-associated factor 6 (TRAF6) autoubiquitination, hence inhibiting NF-κB activation and suppressing IFN-γ production [38–40].

2.5 LAG3 regulations and signaling

LAG3 inhibits CD4-dependent T cell activity by binding to MHC-II due to being structurally homologous with four extracellular immunoglobulin superfamily-like domains. Other investigations demonstrated that LAG3's inhibitory activity is not dependent on CD4 competition, but rather LAG-3 inhibited T cells responding to stable peptide-MHC-II by transducing inhibitory signals via its intracellular domain. Thus, LAG-3 may act more selectively, allowing tolerance to dominant autoantigens to persist [41, 42]. Alternatively, LAG3 can interact with another ligand like Galectin3 in TME and mediate the suppression of CD8 T cells [43]. Besides, liver sinusoidal endothelial cell lectin (LSECtin) can bind to LAG3 in human melanoma, causing tumor growth by abolishing IFN-γ production and proliferation of tumor-specific T cells [44]. Lastly, fibrinogen-like protein 1 (FGL1) was recently discovered as a novel LAG3 ligand. FGL1 is typically produced in trace amounts into the bloodstream by the liver. However, overexpression of FGL1 has been observed in some human malignancies. Inhibiting the interaction between FGL1 and LAG3 by monoclonal antibodies improves T cells' anticancer activity [45].

The signal transduction mechanism of LAG3 is regulated by two transmembranes, a disintegrin and metalloproteinase domain-containing protein 10 and 17 (ADAM10 and ADAM17)-mediated cleavage. TCR signaling enhances ADAM10 and ADAM17 cleavage activity, releasing sLAG3. The function of sLAG3 remains controversial as some studies consider this does not have a biological process, while the others state that sLAG3 allows effective T cell proliferation and function [46]. Besides, sLAG3 affects monocyte differentiation into macrophages and DCs, which have decreased immunostimulatory capacity [47].

2.6 BTLA regulations and signaling

BTLA and CD160 inhibit T cell activity via the same ligand, herpesvirus entry mediator (HVEM). BTLA-HVEM is an example of crosstalk between two superfamilies in which the ligand is a member of the TNF/TNFR superfamily. However, HVEM interaction with members of the TNF superfamily LIGHT (Lymphotoxins, Inducible, competes with herpes simplex virus (HSV) Glycoprotein D for HVEM, expressed by T cells) produces a co-stimulatory signal on B and T cells. Hence, HVEM may be considered as a molecular switch that enables co-signaling between stimulatory and inhibitory T cells. Additionally, signaling between HVEM and its ligands appears to interact bidirectionally. The cis interaction between BTLA and HVEM inhibits the trans-ligation of HVEM by LIGHT and thus inhibits HVEM stimulatory signaling triggered by LIGHT binding [7, 48, 49].

Regarding the inhibitory signaling of BTLA, it follows the mechanism of PD-1/PD-L1 involving ITIM and ITSM to recruit SHP1/SHP2 [50]. In B-chronic lymphocytic leukemia (B-CLL), both HVEM and BTLA are overexpressed. This co-expression of HVEM and BTLA in CLL cells implies that an unsuccessful autocrine inhibitory loop is triggered. In addition, BTLA is typically downregulated during the development of human $CD8^+$ T cells to effector cells. However, BTLA expression was more significant in melanoma-specific $CD8^+$ T lymphocytes specialized for tumor antigens (TA). Despite effector differentiation, BTLA expression remained persistent, confounding T cell proliferation and IFN-γ production. Thus, BTLA may function similarly to PD1 as a T-cell inhibitory receptor in TME [51].

3. Immune checkpoints dysregulation affecting cancer cells

Recent studies have established that immune checkpoint molecules drive cancer growth via various anticancer strategies. The first one is the overexpression of immune checkpoints in cancer cells, immune cells, or the surrounding environment leading to incapabilities of the tumor-specific immune response. Subsequently, immune checkpoints can interfere with metabolic pathways and deplete nutrients needed by immune cells. Lastly, immune checkpoints cripple cancer-specific immune responses by collaborating with regulatory T cells. This section deliberates each strategy thoroughly to get insight into how to combat those actions.

3.1 Overexpression of immune checkpoints favoring tumor growth

Accumulating evidence showed that several immune checkpoint molecules are overexpressed not only on the surface of cancer cells but also in T cells, Tregs, or even in TME. Here, we thoroughly describe how the immune checkpoint is upregulated and then inhibits antitumor activity. PD-1/PD-L1 are overexpressed on the surface of many cancer cells. Several proinflammatory mediators, which are secreted by activated T cells (types I and II IFN-γ, TNF-α, IL-10, and IL-4) or produced in TME (GM-CSF and VEGF), upregulate PD-L1 expression in the cancer cells resulting in suppression of $PD1^+$ T cells activity. Moreover, cancer cells commonly carry altered PTEN (phosphatase and tension homolog deleted on chromosome ten)—PD-L1 suppressor gene—which may activate the S6K1 gene, resulting in a significant increase in PD-L1 mRNA to polysomes, which promotes PD-L1 mRNA translation and plasma membrane expression [6]. In pancreatic cancer cells, PTEN gene deletion influences PD-L1 expression at the translational level by activating the PI3K/AKT downstream mTOR-S6K1 signaling pathway, thereby increasing PD-L1 production and T lymphocyte apoptosis [52, 53].

Furthermore, amplification and translocation of CD274 on chromosome 9p24.1 have been associated with elevated expression of PD-L1 in Hodgkin's lymphoma, small cell lung cancer (SCLC), non-small cell lung cancer (NSCLC), lymphoma,

Epstein-Barr virus (EBV)-positive gastric cancer, and oral squamous cell carcinoma (OSCC). In SCLC, chromosomal rearrangements produce CD274 amplification without changing the open reading frame. It is found in various organs, but it is most commonly found in activated T and B lymphocyte cells, dendritic cells, monocytes, and other types of TCs. The CD274/PD-L1 gene is highly conserved, with homologs discovered across the vertebrate lineage (from *Danio rerio* to Primates), implying its wide range of functions. The CD274/PD-L1 promoter retains CpG methylation sites in the 5′ untranslated region (UTR) and exon 1, but translation begins in exon 2 [54, 55]. JAK2, which encodes Janus kinase 2, an upstream kinase that controls PD-L1 expression, is also present on chromosome 9p, with a high alternation rate. The JAK family has been shown to contribute to PD-L1 upregulation by raising PD-L1 RNA expression through amplification and mutation. Because of the increased activity of the Janus kinase2 (JAK2) signal transducers and activators of transcription (STAT) signaling pathway, PD-L1 expression rises. DNA double-strand breaks (DSBs) consistently activate STAT signaling via the ataxia-telangiectasia mutant (ATM)/ATM- and ataxia-telangiectasia-related (ATR)/checkpoint kinase 1 (Chk1) kinases, leading to PD-L1 expression increase. Moreover, structural changes in CD274's 30 UTR boost protein production and improve cancer-immune evasion in human malignancies [55].

PD-L1 induction has also been associated with inflammatory stimuli such as IL-1b, IL-4, IL-6, IL-10, IL-12, IL-17, IL-27, tumor necrosis factor-α (TNF-α), and transforming growth factor-β (TGF-β). Among the several soluble inflammatory agents, IFN-ϒ is the most important in promoting PD-L1 expression. IFN-ϒ is a proinflammatory cytokine primarily generated by T and NK cells. IFN-ϒ attaches to its receptor, the interferon-gamma receptor (IFNGR), activating the JAK-STAT signaling pathway via STAT1. As a result, it increases the expression of transcription factors, particularly interferon-responsive factors (IRFs). IRF1 is a critical downstream signaling molecule of STAT1 that causes IFN-induced PD-L1. Other proinflammatory agents, IL-4 and TNF-α, have a synergistic impact on the activation of PD-L1 in renal cell carcinoma (RCC) via activating signaling molecules such as NF-κB, IκB, and STAT6. In dendritic cells and monocytes, blocking PD-L1 was associated with decreased IL-10 levels. Furthermore, IL-10 levels on Tyro3, Axl, and Mer (TAM) were closely connected to PD-L1 expression. In monocyte-derived macrophages, IL-12 upregulates PD-L1 expression, but in THP-1-derived macrophages, it downregulates PD-L1 expression. In monocytes, IL-17 is involved in the induction of PD-L1. IL-17 and TNF-α activate NF-κB signaling in prostate cancer and NF-κB and ERK1/2 in colon cancer, respectively, and upregulate PD-L1 expression. PD-L1 expression in dendritic cells is upregulated by IL-1b and IL-27. Furthermore, IL-27 activates phospho-STAT1 and phospho-STAT3 to enhance PD-L1 expression [54, 56].

Meanwhile, CTLA-4 is often constitutively overexpressed on Tregs and has been demonstrated to alter Tregs-mediated immune control. In multiple myeloma patients, FOXP3 and CTLA-4 genes from bone marrow samples were considerably overexpressed [57]. Another sample from peripheral blood mononuclear cells (PBMC) of breast cancer patients showed significantly higher mRNA expression of FOXP3 and CTLA-4 than healthy individuals [58]. Taken together, these results indicated the pivotal role of CTLA-4 in the accumulation of immunosuppressive Tregs in TME, leading to repression of anti-tumor immunity.

Regarding TIM-3 overexpression, it is induced by cytokine stimulation, especially in NK cells. TIM-3 is also extensively expressed on tumor-infiltrating lymphocytes. Similar to its expression pattern during persistent viral infection, TIM-3 is generally co-expressed with PD-1 and represents the most dysfunctional T cell subgroup. TIM-3 overexpression in human malignancies, particularly on immune cells, might be a predictive biomarker for a range of cancers. TIM-3 expression on CD4+ and CD8+ T lymphocytes was enhanced in individuals with hepatitis B virus-related hepatocellular carcinoma (HCC). TIM-3$^+$T cells were replicative senescent and exhibited senescence-related surface and genomic markers. Furthermore, the quantity of tumor-infiltrating cells in TIM-3$^+$ was inversely linked with HCC patient survival [59].

Furthermore, LAG3 is mainly expressed in activated T and natural killer (NK) cells, and it has been identified as a marker for CD4+ and CD8+ T cell activation. Increased LAG3 expression on T cells was observed in combination with other inhibitory receptors such as PD-1, TIGIT, TIM-3, CD160, and 2B4 under pathological conditions such as chronic inflammation or in TME, resulting in T cell exhaustion and reduced cytokine release. In melanoma and colon cancer, LAG3 expression was identified in tissue-infiltrating lymphocytes and peripheral Tregs, tumor-involved lymph nodes, and inside the tumor tissue itself. LAG3 was found on tumor-infiltrating Tregs in patients with head and neck squamous cell carcinoma and non-small cell lung cancer [60].

Similar to other checkpoint molecules, TIGIT is also significantly expressed on Tregs taken from PBMC of cancer patients, and it is further elevated in the TME. Increased TIGIT expression in Tregs is coupled with hypomethylation and FOXP3 binding at the TIGIT gene, distinguishing Tregs from activated effector CD4$^+$ T cells. Furthermore, the Fap2 protein from *Fusobacterium nucleatum*, an anaerobic Gram commensal bacteria linked to colorectal cancer, binds directly to TIGIT but not CD226 to suppress NK cells and T cell-mediated tumor response. These findings imply that the gut microbiota modulates innate immune responses via TIGIT [61].

3.2 Immune checkpoints mediating metabolic reprogramming in TME

Due to cancer cells' resource intake and vascularization defects, TME is typically deficient in nutrients and oxygen. Cancer cells' increased need for glucose promotes competition in the TME, which has a detrimental effect on surrounding cells, such as immune cells. Immune checkpoint proteins have been shown to modulate the metabolic energetics of tumor cells, TME, and the tumor-specific immune response, resulting in metabolic reprogramming of both cancerous and immune cells. For instance, CD80 (B7–1) activated the mTOR kinase in naïve CD8$^+$ T cells via the PI3K and STAT4 pathways in solid tumors. mTOR signaling is required to promote glycolysis via hypoxia-inducible factor-1α (HIF-1α) and protein synthesis for supporting cancer cell growth. This activation shifts nutrition balance, and cancerous cells outcompete the immune cells, then evading immune surveillance [62, 63].

Because amino acids are the building blocks of proteins, their availability is critical for tumor development. At the same time, immune cells need amino acids to differentiate and perform their effector activities, hence regulating tumor formation. Given this, a greater knowledge of how each cell species use amino acids in the TME looks critical for successfully stimulating anti-tumor immunity. Tryptophan deficiency impairs CD8$^+$ T cell functions and enhances CD4$^+$ Tregs cell functions, resulting in immunosuppression mediated by the CTLA-4 and PD-1/PD-L1 pathways. The effects are achieved mechanistically by activating the stress response kinase GCN2, which inhibits mTORC2 and its downstream target AKT [64, 65]. The other amino acids, such as glutamine and arginine, are also extensively consumed by the tumors and directly impoverish T cells, leading to the development of immunosuppressive TME [66]. Additionally, tumors may produce and accumulate toxic compounds like aerobic glycolysis byproduct (lactate) in TME, leading to local acidification. Lactate acidosis and hypoxia can activate HIF-1α and then upregulate PD-L1, further inhibiting T-cell responses specific to tumors. Besides, an acidic condition in the surrounding tumors environment suppresses cytokine production (IFN-γ) and limits the activity of T cell cytotoxic, NK cells, and dendritic cells [66, 67].

In contrast to the effector T cells, glucose deprivation may exert a negligible effect on intratumoral Tregs and lactic acid found in the TME may offer nourishment, thus supporting the immunosuppressive function of Tregs [68]. In addition, Tregs differentiation and recruitment is also supported by kynurenine, a metabolite produced from tryptophan through indoleamine 2,3 dioxygenase (IDO)-catabolization in TME [69]. Furthermore, hypoxia and fatty acids production may facilitate Tregs accumulation, thereby favoring its suppressive function [70].

3.3 Interaction between immune checkpoints and Treg cells

Another immune checkpoint favoring cancer growth strategy is its interaction with Tregs cells either by the expression on Tregs surface or inducing Tregs population and function. Treg cells function in the immune system to regulate and suppress other effector T cells. These cells are responsible for the homeostatic process of the immune system to maintain its unresponsiveness to self-antigens and protect the body from autoimmune reactions or excessive inflammation [71]. However, in this context, the interaction of two immunosuppressive mechanistics is critical in cancer survival from immunosurveillance and progression.

Almost all of the immune checkpoint molecules discussed in this chapter, except BTLA, are expressed in Tregs [72]. CTLA-4 is expressed constitutively on Tregs and induced on effector T cells when activated. CTLA-4 deficiency in Tregs was shown to affect their suppressive effects in animal models. Upon TCR stimulation, CTLA-4 is constitutively recruited on the Tregs cell surface, allowing continuous transendocytosis signaling. Hence, Tregs ($CD4^+$ $Foxp3^+$) can outperformed activated conventional T cells ($CD4^+$ $Foxp3^-$) [73]. Subsequently, downregulation of B7 ligands on APCs leading to diminished CD28 co-stimulation is another way by which Tregs are hypothesized to govern effector T cells [74, 75].

In tumor tissue of non-small cell lung cancer (NSCLC) patients, the PD-L1 expressing $CD25^+$ $CD4^+$ (PD-$L1^{hi}$Tregs) population is higher than in blood or normal tissue. Interestingly, PD-$L1^{hi}$Tregs also correlated with PD-1^+ CD8 [76]. In another cancer, highly expressed PD-L1 glioblastoma cells can induce Tregs expansion and maintain its immunosuppressives through PD-1/PD-L1 stimulation. Disrupting the PD-L1/PD-1 axis could target two immunosuppressive mechanisms: inhibition of signaling due to PD-1/PD-L1 ligation and stimulatory proliferation of Tregs cells, which indirectly promotes immunoresistance of high PD-L1 cancers. Thus, Tregs abundance may be a predictive biomarker for patients likely to react to anti-PD-1/anti-PD-L1 therapy or monitor treatment response [77].

Multiple immune checkpoints protein can coexpress and accumulate on the T cell surface, thus increasing dysfunctionality. On $CD8^+$ TILs, it is found that TIGIT is coexpressed with TIM-3, PD-1, and LAG-3 [78]. Although, further investigation is needed to show whether these pathways synergize and whether coblockade is becoming a more efficient immunotherapeutic approach.

4. Future challenges and applicability of immune checkpoint inhibitors

Immunological tolerance is normally maintained so that the immune system can recognize and distinguish between self and non-self antigens or neoantigens. Although the immune system is expected to protect the host from exposure to non-self antigens, its robust effector mechanism allows to reverse the attack and disrupt the homeostasis of the immune system. Immune checkpoints, which have gained notoriety as possible cancer therapy targets, are essential immunoregulatory processes found throughout the body. Dysregulation of immune checkpoints promotes tumor cell evasion and plays a significant role in cancer pathogenesis. Therefore, several monoclonal antibodies have been made to block the interaction between ligand and receptor of immune checkpoints, enhancing host immunologic competence against tumors. The list of immune checkpoints inhibitors (ICI), which gained Food and Drug Administration (FDA) approval or are in ongoing clinical trials, is comprehensively summarized in [79]. However, only a tiny proportion of patients respond meaningfully to these therapies due to the signaling complexity and overlapped pathways as mentioned above. Thus, new routes and compounds are being investigated to enhance therapeutic responsiveness and applicability. In clinical practice, the difficulties in treating cancer patients revolve on eliminating the tumor and alleviating symptoms such as pain, fatigue, nausea/vomiting, cough, and diarrhea. Then, concomitant use of medications is negligible and generates new threats for drug interaction such as analgesics [80], steroids [81], antibiotics [82], or many others. Moreover, the use of ICI is often associated with immune-related adverse effects (irAEs). A retrospective study reported that among 1091 patients receiving ICI therapy, 487 (44.63%) patients experienced adverse effects. The most common is fatigue (13.9%), then dermatologic irAEs (12%), endocrine-related irAEs (9.89%), gastrointestinal toxicities (8.4%) and hepatotoxicities (4.94%) [83].

Acknowledgements

The authors thank our institutions Akademi Farmasi Surabaya and the University of Surabaya, for the support.

Conflict of interest

The authors declare no conflict of interest.

Author details

Rahmad Aji Prasetya[1*] and Devyani Diah Wulansari[2]

1 Akademi Farmasi Surabaya, Surabaya, Indonesia

2 University of Surabaya, Surabaya, Indonesia

*Address all correspondence to: prasetya.ra@akfarsurabaya.ac.id

References

[1] Thommen DS, Schumacher TN. T cell dysfunction in cancer. Cancer Cell. 2018;**33**:547-562. DOI: 10.1016/J.CCELL.2018.03.012

[2] Chen DS, Mellman I. Oncology meets immunology: The cancer-immunity cycle. Immunity. 2013;**39**:1-10. DOI: 10.1016/j.immuni.2013.07.012

[3] Takeuchi Y, Nishikawa H. Roles of regulatory T cells in cancer immunity. International Immunology. 2016;**28**: 401-409. DOI: 10.1093/intimm/dxw025

[4] Pardoll DM. The blockade of immune checkpoints in cancer immunotherapy. Nature Reviews Cancer. 2012;**12**:252-264. DOI: 10.1038/nrc3239

[5] Jiang Y, Li Y, Zhu B. T-cell exhaustion in the tumor microenvironment. Cell Death & Disease. 2015;**6**:e1792-e1792. DOI: 10.1038/cddis.2015.162

[6] He J, Hu Y, Hu M, Li B. Development of PD-1/PD-L1 pathway in tumor immune microenvironment and treatment for non-small cell lung cancer. Scientific Reports. 2015;**5**:1-9. DOI: 10.1038/srep13110

[7] He X, Xu C. Immune checkpoint signaling and cancer immunotherapy. Cell Research. 2020;**30**(8):660-669. DOI: 10.1038/s41422-020-0343-4

[8] Wang X, He Q, Shen H, Xia A, Tian W, Yu W, et al. TOX promotes the exhaustion of antitumor CD8 + T cells by preventing PD1 degradation in hepatocellular carcinoma. Journal of Hepatology. 2019;**71**:731-741. DOI: 10.1016/J.JHEP.2019.05.015

[9] Meng X, Liu X, Guo X, Jiang S, Chen T, Hu Z, et al. FBXO38 mediates PD-1 ubiquitination and regulates anti-tumour immunity of T cells. Nature. 2018;**564**:130-135. DOI: 10.1038/s41586-018-0756-0

[10] Zhang J, Bu X, Wang H, Zhu Y, Geng Y, Nihira NT, et al. Cyclin D–CDK4 kinase destabilizes PD-L1 via cullin 3–SPOP to control cancer immune surveillance. Nature. 2018;**553**(7686): 91-95. DOI: 10.1038/nature25015

[11] Li CW, Lim SO, Xia W, Lee HH, Chan LC, Kuo CW, et al. Glycosylation and stabilization of programmed death ligand-1 suppresses T-cell activity. Nature Communications. 2016;**7**(1):1-11. DOI: 10.1038/ncomms12632

[12] Lim SO, Li CW, Xia W, Cha JH, Chan LC, Wu Y, et al. Deubiquitination and stabilization of PD-L1 by CSN5. Cancer Cell. 2016;**30**:925-939. DOI: 10.1016/J.CCELL.2016.10.010

[13] Wang H, Yao H, Li C, Shi H, Lan J, Li Z, et al. HIP1R targets PD-L1 to lysosomal degradation to alter T cell-mediated cytotoxicity. Nature Chemical Biology. 2019;**15**:42-50. DOI: 10.1038/S41589-018-0161-X

[14] Fife BT, Pauken KE, Eagar TN, Obu T, Wu J, Tang Q, et al. Interactions between PD-1 and PD-L1 promote tolerance by blocking the TCR-induced stop signal. Nature Immunology. 2009;**10**:1185-1192. DOI: 10.1038/NI.1790

[15] Marasco M, Berteotti A, Weyershaeuser J, Thorausch N, Sikorska J, Krausze J, et al. Molecular mechanism of SHP2 activation by PD-1 stimulation. Science Advances. 2020;**6**:1-15. DOI: 10.1126/SCIADV.AAY4458

[16] Rota G, Niogret C, Dang AT, Barros CR, Fonta NP, Alfei F, et al.

Shp-2 is dispensable for establishing T cell exhaustion and for PD-1 signaling In vivo. Cell Reports. 2018;**23**:39-49. DOI: 10.1016/J.CELREP.2018.03.026

[17] Hui E, Cheung J, Zhu J, Su X, Taylor MJ, Wallweber HA, et al. T cell costimulatory receptor CD28 is a primary target for PD-1-mediated inhibition. Science. 2017;**355**:1428-1433. DOI: 10.1126/SCIENCE.AAF1292

[18] Seidel JA, Otsuka A, Kabashima K. Anti-PD-1 and anti-CTLA-4 therapies in cancer: Mechanisms of action, efficacy, and limitations. Frontiers in Oncology. 2018;**8**:1-14. DOI: 10.3389/fonc.2018.00086

[19] Rotte A. Combination of CTLA-4 and PD-1 blockers for treatment of cancer. Journal of Experimental and Clinical Cancer Research. 2019;**38**:1-12. DOI: 10.1186/s13046-019-1259-z

[20] Banton MC, Inder KL, Valk E, Rudd CE, Schneider H. Rab8 binding to immune cell-specific adaptor LAX facilitates formation of trans-Golgi network-proximal CTLA-4 vesicles for surface expression. Molecular and Cellular Biology. 2014;**34**:1486-1499. DOI: 10.1128/MCB.01331-13

[21] Qureshi OS, Kaur S, Hou TZ, Jeffery LE, Poulter NS, Briggs Z, et al. Constitutive clathrin-mediated endocytosis of CTLA-4 persists during T cell activation. The Journal of Biological Chemistry. 2012;**287**:9429-9440. DOI: 10.1074/JBC.M111.304329

[22] Lo B, Zhang K, Lu W, Zheng L, Zhang Q, Kanellopoulou C, et al. Patients with LRBA deficiency show CTLA4 loss and immune dysregulation responsive to abatacept therapy. Science. 1979;**2015**(349):436-440. DOI: 10.1126/SCIENCE.AAA1663/SUPPL_FILE/AAA1663-LO-SM.PDF

[23] Mitsuiki N, Schwab C, Grimbacher B. What did we learn from CTLA-4 insufficiency on the human immune system? Immunological Reviews. 2019;**287**:33-49. DOI: 10.1111/imr.12721

[24] Qureshi OS, Zheng Y, Nakamura K, Attridge K, Manzotti C, Schmidt EM, et al. Trans-endocytosis of CD80 and CD86: A molecular basis for the cell-extrinsic function of CTLA-4. Science. 1979;**2011**(332):600-603. DOI: 10.1126/science.1202947

[25] Bio-Rad Laboratories Inc. Role of Immune Checkpoints in Immunolgy. Oxfordshire: Bio-Rad Laboratories, Inc; 2017. pp. 1-8

[26] Chen W, Jin W, Wahl SM. Engagement of cytotoxic T lymphocyte–associated antigen 4 (CTLA-4) induces transforming growth factor β (TGF-β) production by murine CD4+ T cells. Journal of Experimental Medicine. 1998;**188**:1849-1857. DOI: 10.1084/jem.188.10.1849

[27] Mellor AL, Chandler P, Baban B, Hansen AM, Marshall B, Pihkala J, et al. Specific subsets of murine dendritic cells acquire potent T cell regulatory functions following CTLA4-mediated induction of indoleamine 2,3 dioxygenase. International Immunology. 2004;**16**:1391-1401. DOI: 10.1093/INTIMM/DXH140

[28] Huang YH, Zhu C, Kondo Y, Anderson AC, Gandhi A, Russell A, et al. CEACAM1 regulates TIM-3-mediated tolerance and exhaustion. Nature. 2015;**517**:386-390. DOI: 10.1038/NATURE13848

[29] Rangachari M, Zhu C, Sakuishi K, Xiao S, Karman J, Chen A, et al. Bat3 promotes T cell responses and autoimmunity by repressing

Tim-3–mediated cell death and exhaustion. Nature Medicine. 2012;**18**:1394-1400. DOI: 10.1038/NM.2871

[30] Li G, Liang X, Lotze MT. HMGB1: The central cytokine for all lymphoid cells. Frontiers in Immunology. 2013;**4**:68. DOI: 10.3389/FIMMU.2013.00068/BIBTEX

[31] Avery L, Filderman J, Szymczak-Workman AL, Kane LP. Tim-3 co-stimulation promotes short-lived effector T cells, restricts memory precursors, and is dispensable for T cell exhaustion. Proceedings of the National Academy of Sciences of the United States of America. 2018;**115**:2455-2460. DOI: 10.1073/PNAS.1712107115/SUPPL_FILE/PNAS.201712107SI.PDF

[32] Das M, Zhu C, Kuchroo VK. Tim-3 and its role in regulating anti-tumor immunity. Immunological Reviews. 2017;**276**:97-111. DOI: 10.1111/IMR.12520

[33] Sakuishi K, Apetoh L, Sullivan JM, Blazar BR, Kuchroo VK, Anderson AC. Targeting Tim-3 and PD-1 pathways to reverse T cell exhaustion and restore anti-tumor immunity. The Journal of Experimental Medicine. 2010;**207**: 2187-2194. DOI: 10.1084/JEM.20100643

[34] Koyama S, Akbay EA, Li YY, Herter-Sprie GS, Buczkowski KA, Richards WG, et al. Adaptive resistance to therapeutic PD-1 blockade is associated with upregulation of alternative immune checkpoints. Nature Communications. 2016;**7**:1-9. DOI: 10.1038/NCOMMS10501

[35] Zaretsky JM, Garcia-Diaz A, Shin DS, Escuin-Ordinas H, Hugo W, Hu-Lieskovan S, et al. Mutations associated with acquired resistance to PD-1 blockade in melanoma. New England Journal of Medicine. 2016;**375**:819-829. DOI: 10.1056/NEJMOA1604958/SUPPL_FILE/NEJMOA1604958_DISCLOSURES.PDF

[36] Levin SD, Taft DW, Brandt CS, Bucher C, Howard ED, Chadwick EM, et al. Vstm3 is a member of the CD28 family and an important modulator of T-cell function. European Journal of Immunology. 2011;**41**:902-915. DOI: 10.1002/EJI.201041136

[37] Johnston RJ, Comps-Agrar L, Hackney J, Yu X, Huseni M, Yang Y, et al. The immunoreceptor TIGIT regulates antitumor and antiviral CD8(+) T cell effector function. Cancer Cell. 2014;**26**:923-937. DOI: 10.1016/J.CCELL.2014.10.018

[38] Stanietsky N, Rovis TL, Glasner A, Seidel E, Tsukerman P, Yamin R, et al. Mouse TIGIT inhibits NK-cell cytotoxicity upon interaction with PVR. European Journal of Immunology. 2013;**43**:2138-2150. DOI: 10.1002/EJI.201243072

[39] Li M, Xia P, Du Y, Liu S, Huang G, Chen J, et al. T-cell immunoglobulin and ITIM domain (TIGIT) receptor/poliovirus receptor (PVR) ligand engagement suppresses interferon-γ production of natural killer cells via β-arrestin 2-mediated negative signaling. The Journal of Biological Chemistry. 2014;**289**:17647-17657. DOI: 10.1074/JBC.M114.572420

[40] Liu S, Zhang H, Li M, Hu D, Li C, Ge B, et al. Recruitment of Grb2 and SHIP1 by the ITT-like motif of TIGIT suppresses granule polarization and cytotoxicity of NK cells. Cell Death and Differentiation. 2013;**20**:456-464. DOI: 10.1038/CDD.2012.141

[41] Maruhashi T, Okazaki I, Sugiura D, Takahashi S, Maeda TK, Shimizu K, et al. LAG-3 inhibits the activation

of CD4 + T cells that recognize stable pMHCII through its conformation-dependent recognition of pMHCII. Nature Immunology. 2018;**19**:1415-1426. DOI: 10.1038/S41590-018-0217-9

[42] Andrews LP, Marciscano AE, Drake CG, Vignali DAA. LAG3 (CD223) as a cancer immunotherapy target. Immunological Reviews. 2017;**276**:80-96. DOI: 10.1111/IMR.12519

[43] Kouo T, Huang L, Pucsek AB, Cao M, Solt S, Armstrong T, et al. Galectin-3 shapes antitumor immune responses by suppressing CD8+ T cells via LAG-3 and inhibiting expansion of Plasmacytoid dendritic cells. Cancer Immunology Research. 2015;**3**:412-423. DOI: 10.1158/2326-6066.CIR-14-0150

[44] Xu F, Liu J, Liu D, Liu B, Wang M, Hu Z, et al. LSECtin expressed on melanoma cells promotes tumor progression by inhibiting antitumor T-cell responses. Cancer Research. 2014;**74**:3418-3428. DOI: 10.1158/0008-5472.CAN-13-2690

[45] Wang J, Sanmamed MF, Datar I, Su TT, Ji L, Sun J, et al. Fibrinogen-like protein 1 is a major immune inhibitory ligand of LAG-3. Cell. 2019;**176**:334-347.e12. DOI: 10.1016/J.CELL.2018.11.010

[46] Li N, Wang Y, Forbes K, Vignali KM, Heale BS, Saftig P, et al. Metalloproteases regulate T-cell proliferation and effector function via LAG-3. The EMBO Journal. 2007;**26**:494-504. DOI: 10.1038/SJ.EMBOJ.7601520

[47] Buisson S, Triebel F. LAG-3 (CD223) reduces macrophage and dendritic cell differentiation from monocyte precursors. Immunology. 2005;**114**:369-374. DOI: 10.1111/J.1365-2567.2004.02087.X

[48] Sedy JR, Gavrieli M, Potter KG, Hurchla MA, Lindsley RC, Hildner K, et al. B and T lymphocyte attenuator regulates T cell activation through interaction with herpesvirus entry mediator. Nature Immunology. 2005;**6**:90-98. DOI: 10.1038/NI1144

[49] Pasero C, Olive D. Interfering with coinhibitory molecules: BTLA/HVEM as new targets to enhance anti-tumor immunity. Immunology Letters. 2013;**151**:71-75. DOI: 10.1016/J.IMLET.2013.01.008

[50] Watanabe N, Gavrieli M, Sedy JR, Yang J, Fallarino F, Loftin SK, et al. BTLA is a lymphocyte inhibitory receptor with similarities to CTLA-4 and PD-1. Nature Immunology. 2003;**4**:670-679. DOI: 10.1038/NI944

[51] Derré L, Rivals JP, Jandus C, Pastor S, Rimoldi D, Romero P, et al. BTLA mediates inhibition of human tumor-specific CD8+ T cells that can be partially reversed by vaccination. The Journal of Clinical Investigation. 2010;**120**:157-167. DOI: 10.1172/JCI40070

[52] Palma M, Gentilcore G, Heimersson K, Mozaffari F, Näsman-Glaser B, Young E, et al. T cells in chronic lymphocytic leukemia display dysregulated expression of immune checkpoints and activation markers. Haematologica. 2017;**102**:562-572. DOI: 10.3324/haematol.2016.151100

[53] Parsa AT, Waldron JS, Panner A, Crane CA, Parney IF, Barry JJ, et al. Loss of tumor suppressor PTEN function increases B7-H1 expression and immunoresistance in glioma. Nature Medicine. 2006;**13**(1):84-88. DOI: 10.1038/nm1517

[54] Zhang H, Dai Z, Wu W, Wang Z, Zhang N, Zhang L, et al. Regulatory mechanisms of immune checkpoints PD-L1 and CTLA-4 in cancer. Journal of Experimental & Clinical Cancer Research. 2021;**40**:1-22

[55] Fabrizio FP, Trombetta D, Rossi A, Sparaneo A, Castellana S, Muscarella LA. Gene code CD274/PD-L1: From molecular basis toward cancer immunotherapy. Therapeutic Advances in Medical Oncology. 2018;**10**:1-18. DOI: 10.1177/1758835918815598

[56] Garcia-Diaz A, Shin DS, Moreno BH, Saco J, Escuin-Ordinas H, Rodriguez GA, et al. Interferon receptor signaling pathways regulating PD-L1 and PD-L2 expression. Cell Reports. 2017;**19**:1189-1201. DOI: 10.1016/j.celrep.2017.04.031

[57] Braga WMT, da Silva BR, de Carvalho AC, Maekawa YH, Bortoluzzo AB, Rizzatti EG, et al. FOXP3 and CTLA4 overexpression in multiple myeloma bone marrow as a sign of accumulation of CD4+ T regulatory cells. Cancer Immunology, Immunotherapy. 2014 [Internet] [cited 2022 Apr 17];**63**:1189-1197. DOI: 10.1007/S00262-014-1589-9/TABLES/2

[58] Khalife E, Khodadadi A, Talaeizadeh A, Rahimian L, Nemati M, Jafarzadeh A. Overexpression of regulatory T cell-related markers (FOXP3, CTLA-4 and GITR) by peripheral blood mononuclear cells from patients with breast cancer. Asian Pacific Journal of Cancer Prevention : APJCP. 2018 [Internet] [cited 2022 Apr 17];**19**:3019. DOI: 10.31557/APJCP.2018.19.11.3019

[59] Tian T, Li Z. Targeting Tim-3 in cancer with resistance to PD-1/PD-L1 blockade. Frontiers in Oncology. 2021;**11**:3877. DOI: 10.3389/FONC.2021.731175/BIBTEX

[60] Puhr HC, Ilhan-Mutlu A. New emerging targets in cancer immunotherapy: The role of LAG3. ESMO Open. 2019;**4**:e000482. DOI: 10.1136/ESMOOPEN-2018-000482

[61] Chauvin JM, Zarour HM. TIGIT in cancer immunotherapy. Journal for ImmunoTherapy of Cancer. 2020 [Internet] [cited 2022 Apr 18];**8**:e000957. DOI: 10.1136/JITC-2020-000957

[62] Lim S, Phillips JB, Da Silva LM, Zhou M, Fodstad O, Owen LB, et al. Interplay between immune checkpoint proteins and cellular metabolism. Cancer Research. 2017;**77**:1245-1249. DOI: 10.1158/0008-5472.CAN-16-1647/660808/P/INTERPLAY-BETWEEN-IMMUNE-CHECKPOINT-PROTEINS-AND

[63] Rao RR, Li Q, Odunsi K, Shrikant PA. The mTOR kinase determines effector versus memory CD8+ T cell fate by regulating the expression of transcription factors T-bet and Eomesodermin. Immunity. 2010;**32**:67-78. DOI: 10.1016/J.IMMUNI.2009.10.010

[64] Sharma MD, Shinde R, McGaha TL, Huang L, Holmgaard RB, Wolchok JD, et al. The PTEN pathway in Tregs is a critical driver of the suppressive tumor microenvironment. Science Advances. 2015;**1**:1-15. DOI: 10.1126/SCIADV.1500845

[65] Munn DH, Sharma MD, Baban B, Harding HP, Zhang Y, Ron D, et al. GCN2 kinase in T cells mediates proliferative arrest and anergy induction in response to indoleamine 2,3-dioxygenase. Immunity. 2005;**22**:633-642. DOI: 10.1016/J.IMMUNI.2005.03.013

[66] Cerezo M, Rocchi S. Cancer cell metabolic reprogramming: A keystone for the response to immunotherapy. Cell Death & Disease. 2020;**11**(11):1-10. DOI: 10.1038/s41419-020-03175-5

[67] Barsoum IB, Smallwood CA, Siemens DR, Graham CH. A mechanism of hypoxia-mediated escape from

adaptive immunity in cancer cells. Cancer Research. 2014;**74**:665-674. DOI: 10.1158/0008-5472.CAN-13-0992

[68] Angelin A, Gil-de-Gómez L, Dahiya S, Jiao J, Guo L, Levine MH, et al. Foxp3 reprograms T cell metabolism to function in low-glucose, high-lactate environments. Cell Metabolism. 2017 [Internet] [cited 2022 Apr 17;**25**:1282-1293.e7. DOI: 10.1016/J.CMET.2016.12.018

[69] Wainwright DA, Balyasnikova IV, Chang AL, Ahmed AU, Moon KS, Auffinger B, et al. IDO expression in brain tumors increases the recruitment of regulatory T cells and negatively impacts survival. Clinical Cancer Research. 2012 [Internet] [cited 2022 Apr 17];**18**:6110. DOI: 10.1158/1078-0432.CCR-12-2130

[70] Wang H, Franco F, Ho PC. Metabolic regulation of Tregs in cancer: Opportunities for immunotherapy. Trends Cancer. 2017 [Internet] [cited 2022 Apr 17;**3**:583-592. DOI: 10.1016/J.TRECAN.2017.06.005

[71] Pacella I, Piconese S. Immunometabolic checkpoints of treg dynamics: Adaptation to microenvironmental opportunities and challenges. Frontiers in Immunology. 2019;**10**:1-17. DOI: 10.3389/fimmu.2019.01889

[72] Sasidharan Nair V, Elkord E. Immune checkpoint inhibitors in cancer therapy: A focus on T-regulatory cells. Immunology and Cell Biology. 2018;**96**:21-33. DOI: 10.1111/IMCB.1003

[73] Ovcinnikovs V, Ross EM, Petersone L, Edner NM, Heuts F, Ntavli E, et al. CTLA-4-mediated transendocytosis of costimulatory molecules primarily targets migratory dendritic cells. Science Immunology. 2019;**4**:1-12. DOI: 10.1126/SCIIMMUNOL.AAW0902

[74] Buchbinder EI, Desai A. CTLA-4 and PD-1 pathways similarities, differences, and implications of their inhibition. American Journal of Clinical Oncology: Cancer Clinical Trials. 2016;**39**:98-106. DOI: 10.1097/COC.0000000000000239

[75] Saleh R, Elkord E. Treg-mediated acquired resistance to immune checkpoint inhibitors. Cancer Letters. 2019;**457**:168-179. DOI: 10.1016/j.canlet.2019.05.003

[76] Wu SP, Liao RQ, Tu HY, Wang WJ, Dong ZY, Huang SM, et al. Stromal PD-L1–positive regulatory T cells and PD-1–positive CD8-positive T cells define the response of different subsets of non–small cell lung cancer to PD-1/PD-L1 blockade immunotherapy. Journal of Thoracic Oncology. 2018;**13**:521-532. DOI: 10.1016/J.JTHO.2017.11.132

[77] DiDomenico J, Lamano JB, Oyon D, Li Y, Veliceasa D, Kaur G, et al. The immune checkpoint protein PD-L1 induces and maintains regulatory T cells in glioblastoma. OncoImmunology. 2018;**7**:1-11. DOI: 10.1080/2162402X.2018.1448329/

[78] Kurtulus S, Sakuishi K, Ngiow SF, Joller N, Tan DJ, Teng MWL, et al. TIGIT predominantly regulates the immune response via regulatory T cells. The Journal of Clinical Investigation. 2015;**125**:4053-4062. DOI: 10.1172/JCI81187

[79] Marin-Acevedo JA, Kimbrough EO, Lou Y. Next generation of immune checkpoint inhibitors and beyond. Journal of Hematology & Oncology. 2021;**14**:45. DOI: 10.1186/s13045-021-01056-8

[80] Prasetya RA, Metselaar-Albers M, Engels F. Concomitant use of analgesics and immune checkpoint inhibitors in non-small cell lung cancer: A

pharmacodynamics perspective. European Journal of Pharmacology. 2021;**906**:1-8. DOI: 10.1016/j.ejphar.2021.174284

[81] De Giglio A, Mezquita L, Auclin E, Blanc-Durand F, Riudavets M, Caramella C, et al. Impact of Intercurrent introduction of steroids on clinical outcomes in advanced non-small-cell lung cancer (NSCLC) patients under immune-checkpoint inhibitors (ICI). Cancers. 2020;**12**:2827. DOI: 10.3390/CANCERS12102827

[82] Iglesias-Santamaría A. Impact of antibiotic use and other concomitant medications on the efficacy of immune checkpoint inhibitors in patients with advanced cancer. Clinical and Translational Oncology. 2020;**22**:1481-1490. DOI: 10.1007/s12094-019-02282-w

[83] Thapa B, Roopkumar J, Kim AS, Gervaso L, Patil PD, Calabrese C, et al. Incidence and clinical pattern of immune related adverse effects (irAE) due to immune checkpoint inhibitors (ICI). Journal of Clinical Oncology. 2019;**37**:e14151-e14151. DOI: 10.1200/JCO.2019.37.15_suppl.e14151

Chapter 4

Pattern Recognition Receptor-Mediated Regulatory T Cell Functions in Diseases

Ece Oylumlu, Goksu Uzel, Lubeyne Durmus, Meric Tas, Damla Gunes and Ceren Ciraci

Abstract

The advent of new technologies in gene expression, immunology, molecular biology, and computational modeling studies has expedited the discovery process and provided us with a holistic view of host immune responses that are highly regulated. The regulatory mechanisms of the immune system lie not only in weakening the attacker directly but also in fortifying the defender for the development of an efficient adaptive immune response. This chapter reviews a comprehensive set of experimental and bioinformatic studies designed to deepen the current knowledge on the regulatory T cells (Tregs) in the context of Pattern Recognition Receptors (PRRs). Initially, we examined both membrane-bound Toll-like Receptors (TLRs) and C Type Lectin Receptors (CLRs); and cytosolic NOD-like Receptors (NLRs) and RIG-I like Receptors (RLRs) in Tregs. Then, we revisited the disease conditions associated with regulatory T cells by emphasizing the essential roles of PRRs. Expanding our knowledge and strategies on the regulatory mechanisms are likely to provide our best chances for long-term disease control and maintenance of homeostasis.

Keywords: pattern recognition receptors, regulatory T cells, NLRs, TLRs, CTLs, RLRs, disease

1. Introduction

Regulatory T cells (Tregs) are a subtype of T cells that are responsible for the maintenance of homeostasis and tolerance to self-molecules. They mediate their action by suppressing the T cell proliferation, and cytokine productions; thereby preventing autoimmunity [1]. In this sense, Tregs can be both helpful to the host by alleviating the immunopathology, and immune system related tissue damages and unfortunately can also be harmful to the host by sabotaging the properly induced immune responses against pathogens [2]. Therefore, harnessing Treg mechanisms could be an efficient therapeutic approach to treat some distinct diseases, including infectious diseases, asthma and allergies, and cancer [3].

It is established that immune cells rely on the germ-line encoded pattern recognition receptors (PRR) that recognize common structural motifs shared by pathogens called pathogen-associated molecular patterns (PAMPs), and also recognize cellular stress and death via molecules known as damage-associated molecular patterns (DAMPs) to initiate inflammation and activate tissue repair mechanisms [4]. As much as the efficacy of PRRs in executing an immune response is critical for the host, it can also be the reason for unintended responses. Fortunately, understanding of how PRRs drive these responses has expanded enormously in the last few decades. In this chapter, we compile the available data using the open-source databases and the current knowledge in an attempt to discern the layers of complex mechanisms from a regulatory T cell standpoint.

2. The expression profiles of Pattern Recognition Receptors (PRRs) in regulatory T cells (Tregs)

2.1 Membrane-bound Toll-Like Receptors (TLRs)

Toll-like receptors (TLRs) have critical roles in the initial defense of innate and adaptive immunity [5]. TLRs which are type I integral membrane receptors have three domains: The N-terminal domain (NTD), which is located either on the outside of the cell membrane or in endosomes, a single helix transmembrane domain that is in the center, and the C-terminal domain (CTD), which is located in the cytoplasm. The N-terminal ectodomains contain a conserved 19–25 tandem leucine-rich repeat (LRR) region leading to the recognition of PAMPs and DAMPs. NTD also contains glycan moieties to bind ligands from different pathogens. On the other hand, the CTD contains the toll-IL-1 receptor (TIR) homologous domain, which enables the interaction with downstream adaptor proteins for signal transduction and thus, activation of the signaling pathway [6–8]. To date, 13 members of the mammalian TLRs have been identified. 10 members of this receptor family are expressed in humans (TLR1–10), while 12 members are expressed in mice (TLR1–9, TLR11–13). Each TLR can recognize different PAMPs from various pathogens. TLRs divided into two classes according to their localization: cell surface and intracellular [4, 5]. TLR1/TLR2 (triacyl lipopeptides), TLR4 (lipopolysaccharide), TLR5 (flagellin), TLR2/TLR6 (lipoproteins), TLR10 (bacterial 23S rRNA) are expressed in the cell membrane and TLR3 (dsRNA), TLR7/TLR8 (ssRNA), TLR9 (unmethylated CpG DNA), TLR11 (flagellin or profilin-like molecule from *T. gondii*), TLR12 (profilin from *T. gondii*), and TLR13 (bacterial 23S rRNA) are expressed in endosomal membranes [4]. TLRs are important for the proper functioning of Tregs because they directly mediate the pathogen sensing. Tregs have higher expression levels of TLR4, TLR5, TLR7, and TLR8 as compared to the effector T cells in humans; however, especially TLR2, TLR4, TLR5, and TLR8 activation has different effects on the differentiation, expansion, and proliferation of Tregs [9]. Although their numbers increased, Tregs lost their suppressive function when treated with PAM3Cys to activate TLR2 signaling, along with T cell receptor (TCR) and interleukin 2 (IL2) stimulation in order for Treg differentiation and function [10, 11]. In this study, the immune response was suppressed neither *in vitro* nor *in vivo* in mice who underwent acute infection, as Tregs were not activated by the induction of the TLR2 signaling pathway [9]. However, when the TLR2 ligand was removed, Tregs' suppressive functions were recovered [11]. MyD88 is an adaptor protein located downstream of TLR2 signaling. Additionally, the effects of TLR2-MyD88 signaling pathway were examined in Tregs isolated from MyD88 deficient mice, and a reduction in suppressive functions of Tregs in the absence of MyD88 was reported [5, 9]. Suppressive functions of Tregs were induced by cell-contact mechanisms and secretion of TGFβ and IL10, which are immunosuppressive cytokines, without an increase in the number of Tregs [11].

Unlike TLR2, LPS indued TLR4 enhanced the suppression effect owing to the increase in FOXP3 expression in both human and murine Tregs. In fact, LPS not only increased the number of Tregs, but also increased the expression of activation markers in cells [12]. As for TLR5, the suppression capacity of CD4 + CD25+ Tregs increased as a result of TLR5 activation by flagellin [13]. TLR8, on the other hand, has been shown to abundantly express in Tregs and upon stimulation with TLR8 ligand, the suppressive function of Tregs was abolished but it had no effect on the Tregs proliferation [14]. Lastly, studies have revealed that TLR9 ligand CpG oligodeoxy-nucleotide induced proliferation of both effector T cells (Teff) and Tregs and partly inhibits the suppressive activity of regulatory T cells in rats. This combined effect of TLR9 ligand is likely to reinforce the adaptive immunity by not only expanding effector cells but also by mitigating the suppressive activity of regulatory T cells [15]. Taken together, multiple studies suggest that the suppressive properties of regulatory T cells with respect to their proliferation and cytokine production capacities may differ depending on the induction and the differential expression of different TLRs in regulatory T cells (**Figure 1**).

By examining the open-source databases of immune cell-specific gene expression profiles, we evaluated the results from published literature for TLRs in Tregs and compared their expressions among all available immune cell types. TLRs did not display a Treg specific high expression across the datasets we examined [16–18]. The DICE database generated by Schmiedel *et al* contains *RNA-seq* data of 13 immune cells including naive and memory Tregs (and two ex vivo activated cell types) collected from peripheral blood mononuclear cell fraction of 91 healthy human donors. We used this dataset for a detailed search in Tregs. Expression patterns of TLRs in Tregs are shown as boxplots (**Figure 1A** and **B**). Although TLRs had relatively low expression profiles, TLR1, TLR2, TLR5, and TLR6 are the ones with the most prominent representation and differential expression in Tregs (**Figure 1C** and **Table 1**). Interestingly, TLR1, TLR2, and TLR5 had higher expression in naive Tregs than memory Tregs, whereas TLR6 expression was higher in memory Tregs than naïve Tregs. Additionally, this dataset presents the sex-biased transcripts for immune cell types. For example, TLR1 in naive Tregs were revealed as one of genes having female bias [16].

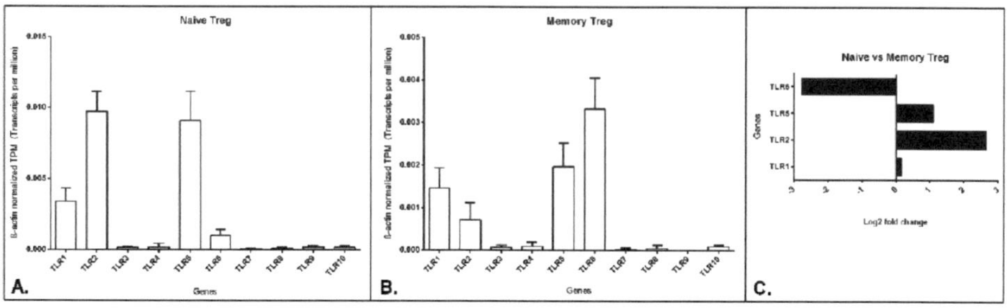

Figure 1.
TLR expression profile in naive Tregs and memory Tregs: Expression of TLRs in A: naive (CD3+, CD4+, CD25 high, CD45RA+, CD127low); B: memory (CD3+, CD4+, CD25 high, CD45RA-, CD127low) regulatory T cells. C: Differentially expressed genes in between two Treg populations. Expression of genes reported as TPM was normalized to β-actin (TPMgene/TPMβ-actin) for individual representation of naive and memory Tregs. Results are depicted as bar plots in GraphPad Prism.

	Biotype	Naive treg mean expression (TPM)	Memory treg mean expression (TPM)	Log 2 fold change	Adjusted p-value
TLR1	Protein coding	7.18	6.1	0.16	0.017
TLR2	Protein coding	14.71	2.27	2.66	6.90E−233
TLR5	Protein coding	15.46	6.59	1.12	6.90E−35
TLR6	Protein coding	1.08	6.31	−2.78	6.50E−251

Table 1.
Differentially expressed toll-like receptor family members in naïve and memory tregs, p-values (https://dice-database.org/).

2.2 Membrane-bound C-type lectin (CTLs) receptors in tregs

C-type lectin receptors (CLRs) also belong to the family of pattern recognition receptors (PRRs) which recognize PAMPs and induce innate immune responses [19]. CLRs comprise a variety of receptors including selectins, collectins, proteoglycans, and lymphocyte lectins. This receptor family possesses at least one structurally homologous carbohydrate recognition domain (CRD), also known as C-type lectin-like domain (CTLD), that determines the carbohydrate specificity [20]. Based on the protein location site on the cell membrane, CLRs are categorized as transmembrane receptors and secretory receptors [21]. Upon ligand recognition by CLRs, most of them are able to induce intracellular pathways and caspase-recruitment domain-containing domain protein 9 (CARD9) pathway, which are vital, and their dysregulation or malfunctioning may result in critical infections in humans and mice [22, 23].

CLRs are expressed on antigen presenting cells (APCs), such as dendritic cells (DCs) and macrophages, and play essential roles in antigen uptake and presentation. In this regard, they are divided mainly into two subgroups: type I and type II CLRs. Two subsets of transmembrane CLRs can be classified based on their CRDs; type I and II. Type I CLRs are mannose receptor family (MR) and DEC-205, whereas type II CLRs are sialoglycoprotein receptor family, DC-associated C-type lectin 1 (dectin-1) and macrophage galactose C type lectin (MGL) [24]. MGLs are able to recognize particularly terminal α and β N-acetylglactoseamine (GalNAc or Tn) residues from filovirus, helminths, bacteria, and tumor-associated antigens in humans [25]. Human counterparts of MGL in mice are MGL1 and MGL2, which are expressed on DCs [26] and activated macrophages [27]. The potency of human MGL was shown by Napoletano *et al.* as an adjuvant for designing novel anticancer vaccines because MGL engagement led to the increased antigen presenting potential in DCs and enhancement in antigen-specific $CD8^+$ cell activation [28]. Dectin-1 is another type II C-type lectin receptor, which is involved in the antifungal immunity by recognizing β-1,3-glucans in the cell wall of several pathogenic fungi. Dectin-1 is able to induce several responses including phagocytosis, through spleen tyrosine kinase (Syk)/CARD9 pathway, which results in the cytokine production [29, 30]. Besides innate immunity, dectin-1 is also capable of triggering adaptive immune responses. For instance, curdlan activated dectin-1 in DCs, skewed the T cell polarization into Th17 and Th1

CD4$^+$ T cell subsets in mice *in vitro* [31]. Finally, C-type lectin receptor CD69 have been shown to control T cell development and homeostasis in mice along with miR155 as both CD69 and miRNA155 were simultaneously regulated to ensure a balanced immunity [32].

Due to lack of studies focusing on CLRs in Treg populations, we used the DICE database to evaluate the expression profiles of CLRs in naive and memory Tregs collected from PBMC fractions of healthy donors. As described in **Figure 2A** and **B**, among all CLRs with low expressions, CLEC4A (DICR) is the one with higher representation in both cell types. Even though CLEC7A (Dectin 1) had low expression in both cell types, it was the only differentially expressed gene (**Figure 2C** and **Table 2**). Taken together, CLRs did not have a notable expression profile in Tregs, thus our analysis is in agreement with previously reported data [16–18].

2.3 Cytosolic NOD-like receptors (NLRs) in regulatory T cells

Nucleotide binding oligomerization domain (NOD)–like receptors (NLRs) are a family of cytoplasmic PRRs that are known to drive the initial innate immune responses. There are 22 NLR members in human and 34 in mice [33], and they are characterized by a C-terminal domain of leucine rich repeats (LRRs) which senses PAMPs and danger molecules (DAMPs); a central NACHT domain that facilitates NLR oligomerization; and an N-terminal signaling domain [34]. The NLR members have been classified in 5 subfamilies based on their N-terminal domain: i) NLRA (CIITA), which contains acidic transactivation domain; ii) NLRB (NAIP) subfamily having an N-terminal baculovirus inhibition of apoptosis repeat (BIR) domain; iii) NLRC subfamily that contains caspase activation and recruitment domain (CARD) and allows direct interaction of NLR family members; iv) NLRP subfamily that bears a pyrin domain (PYD); and v) NLRX subfamily that has a mitochondria-targeting sequence required for its trafficking [34]. Some of the NLR members including NLRP1, NLRP3, NLRP6, NLRP7, NLRP12, and NLRC4 are reported to assemble large multimeric protein complexes called "Inflammasomes" which regulate the activation of caspases-1 [35, 36]. The signaling pathway where the assembled inflammasome activates pro-caspase-1 into its catalytically active form is generally

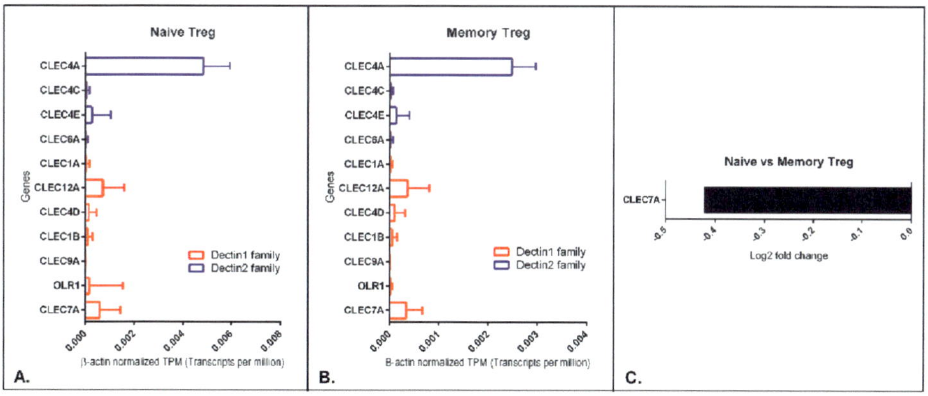

Figure 2.
C type lectin expression in naive Tregs and memory Tregs: Expression of CLRs in A: naive (CD3+, CD4+, CD25 high, CD45RA+, CD127low); B: memory (CD3+, CD4+, CD25 high, CD45RA-, CD127low) regulatory T cells. C: Expression of genes reported as TPM was normalized to β-actin (TPMgene/TPMβ-actin) for individual representation of naive and memory Tregs. Results are depicted as bar plots in GraphPad Prism.

	Biotype	Naive treg mean expression (TPM)	Memory treg mean expression (TPM)	Log 2 fold change	Adjusted p-value
CLEC7A	Protein coding	2.05	2.44	−0.42	0.031

Table 2.
Differentially expressed C type lectin receptor family members in naïve and memory tregs, p-values. (https://dice-database.org/).

referred to as the canonical inflammasome whose activation requires two steps: transcription and oligomerization. The first step is regulated by innate immune signaling, primarily by TLR signaling, and/or cytokine receptors such as TNF which leads to the production of biologically inactive pro interleukin-1β (IL1β), IL18, and NLR transcription via nuclear factor-κB (NF-κB) activation. The second step leads to inflammasome oligomerization and eventually caspase-1 activation which, in turn, results in IL1β and IL18 processing and secretion [37]. Biologically active IL1β and IL18 promote inflammatory and antimicrobial responses and activate different helper T cell subsets such as Th1 and Th17 cells [38]. Although NLRs activation leads to numerous signaling cascades which subsequently initiate the appropriate immune responses including the regulation of B and T cell functions [39], studies focusing on NLRs especially in Tregs are limited. Hence, utilizing open-source datasets, we evaluated the expression patterns of NLRs in Treg populations. Firstly, to address whether there is a Treg specific NLR expression, we compared 28 and 29 cell types studied by Ota *et al.*, and Monaco *et al.*, respectively and showed that there is no NLR specifically expressed in Tregs [17, 18]. Based on the current database, NLRC5 and NLRP1 appear to have higher expression levels in memory Tregs among 13 immune cells included by the DICE database. Next, we examined the data obtained from this database for NLR expression by focusing on Tregs separately (**Figure 3**). As detailed in **Figure 3A** and **B**, most NLR family members have low expressions in both I and memory Treg populations. NLRC3, NLRC5, and NLRP1 have higher expression levels than the rest of the NLRs in both cells. Interestingly, Schmiedel *et al.* listed NLRP2 transcripts as sex-biased (toward females) in both Treg populations [16]. Several NLRs are detected to be differentially expressed naive and memory Tregs are compared with one another (**Figure 3C** and **Table 3**). CIITA, NOD2, NLRC5 expressions were significantlyIr in naive Tregs, while NLRP6 has a relatively higher differential expression profile than the remaining NLRs in memory Tregs. Critical roles, if any, of NLRs' expressional diversity within and in between Treg populations might need further investigation.

Apart from these, studies concentrating on NLRP3 and NOD2 roles in directing (Treg) differentiation and function demonstrated that NLRP3 negatively regulates Treg differentiation in an inflammasome-independent manner via translocation to the nucleus and subsequently interacting with Kpna2 [40]. Of note, immunoprotective roles were reported for NLRP3 inflammasome in controlling the Th1/Th17 immunity against fungal infection of pulmonary paracoccidioidomycosis by suppressing the expansion and migration of Tregs in mice [41]. In addition to NLRP3, NOD2 has been shown to get activated by muramyl dipeptide (MDP), resulting in NF-κB translocation to nucleus in primary human FOXP3+ T cells thereby protecting from death receptor Fas-mediated apoptosis [42]. Finally, MDP-stimulated migration of Tregs has been shown to suppress the Th17 cells in the lungs of influenza A virus-infected

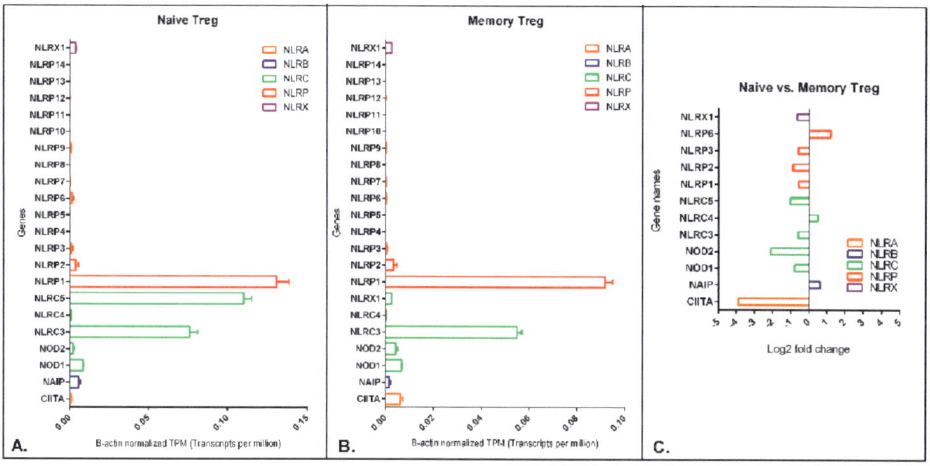

Figure 3.
NLR expression profile in naive Tregs and memory Tregs: Expression of NLRs in A: naive (CD3+, CD4+, CD25 high, CD45RA+, CD127low); B: memory (CD3+, CD4+, CD25 high, CD45RA-, CD127low) regulatory T cells. C: Expression of genes reported as TPM was normalized to β-actin (TPMgene/TPMβ-actin) for individual representation of naive and memory Tregs. Results are depicted as bar plots in GraphPad Prism.

mice [43]. Together, these results suggest that the control of inflammation during fungal and viral infections is mediated by Tregs, with the contribution of NLRs.

2.4 Cytosolic retinoic acid-inducible Gene (RIG-i) like receptor

Type I interferons are proven to be indispensable during viral infections for their ability to generate an antiviral state. Although they are expressed at low levels, they are induced during the course of an infection which is detected by the presence of the foreign nucleic acids [44]. One example for such sensors is retinoic acid-inducible gene (RIG) I like receptors (RLR) whose activation results in type I interferons [45, 46]. RLRs sense viral RNA in the cytosol. The RLR family comprises three proteins: i) RIG-I; ii) melanoma differentiation-associated antigen 5 (MDA-5); and iii) laboratory of genetics and physiology 2 (LGP2) [47]. All these RLRs share a common structure including a central helicase domain responsible for ATP hydrolysis to unwind RNA and a C-terminal regulatory (CTR) domain adjacent to the helicase core. The CTR domain aids to distinguish the self RNAs from the foreign RNA fragments within the cellular environment. Added to these domains, RIG-I and MDA-5 have N-terminal caspase activation and recruitment domains (CARDs) that are required for downstream signaling through the interaction with CARDs of CARD containing adaptor proteins. Dissimilar to RIG-I and MDA5, LGP2 lacks the CARD domain. Instead, LGP2 is of importance to regulate the RIG-I and MDA5 directed antiviral responses [48–50].

To investigate the RLR expressions in Treg cells, we used the expression data from the DICE database (**Figure 4**). Similar to other PRRs, we did not observe a Treg specific expression of RLRs. As depicted in **Figure 4A** and **B,** RLRs have similar expression patterns among their members, DDX58 (RIG1) having a relatively higher expression trend as compared to IFIH1 (MDA5) andI8 in both naive and memory Tregs. Next, we listed the differentially expressed genes (**Figure 4C** and **Table 4**). Interestingly, Schmiedel *et al.* identified IFIH1 (MDA5) transcripts to have

	Biotype	Naive TREG mean expression (TPM)	Memory treg mean expression (TPM)	Log 2 fold change	Adjusted p-value
CIITA	Protein coding	1.6	16.65	−3.89	8.50E−262
NAIP	Protein coding	20.72	13.32	0.62	1.10E−09
NOD1	Protein coding	16.61	26.42	−0.79	2.90E−116
NOD2	Protein coding	3.1	11.2	−2.1	5.90E−169
NLRC3	Protein coding	168	237.03	−0.61	7.40E−50
NLRC4	Protein coding	1.39	0.96	0.5	4.40E−09
NLRC5	Protein coding	222.01	418.41	−1.03	1.50E−81
NLRP1	Protein coding	272.66	376.61	−0.57	6.90E−54
NLRP2	Protein coding	8.35	14.42	−0.88	7.40E−08
NLRP3	Protein coding	1.33	1.81	−0.59	7.00E−08
NLRP6	Protein coding	3.59	1.33	1.21	5.90E−17
NLRX1	Protein coding	6.78	9.8	−0.65	2.20E−19

Table 3.
Differentially expressed NOD like receptor family members in naïve and memory tregs, p-values (https://dice-database.org/).

female-biased sex ratios. Data analysis indicates that all RLRs are expressed in Tregs from healthy donors [16].

RLRs, MDA-5, and RIG-I are ubiquitously expressed in the cytoplasm of immune cells including Tregs. Although exhaustion of Tregs following bacterial ligand treatments has been demonstrated [14, 51, 52], the impact of viral infection on Treg derived suppression remained elusive. A study by Anz *et al.* suggested a direct suppression mechanism through the activation of RLRs. In this particular study, regulatory and effector T cells from wild-type and MDA-5 deficient mice were cocultured and infected with encephalomyocarditis virus. Results suggested that MDA-5 deficient Tregs lost their ability to suppress the immune responses when compared with wild type counterparts during the viral infection [53].

Because IFN-beta promoter stimulator (IPS-1) is the main adaptor protein of RLR signaling [54], its influence on the RLR signaling during West Nile Virus (WNV) infection were studied with respect to Tregs using IPS-1 deficient mice. Conceivably, uncontrolled inflammatory responses including the more pronounced immune cell responses and failure in virus neutralization were identified with the lack of Tregs expansion which is a characteristic of WNV infection [55]. Moreover, Xu *et al.* showed the intrinsic suppression ability of RNA stimulated RIG-I in Treg

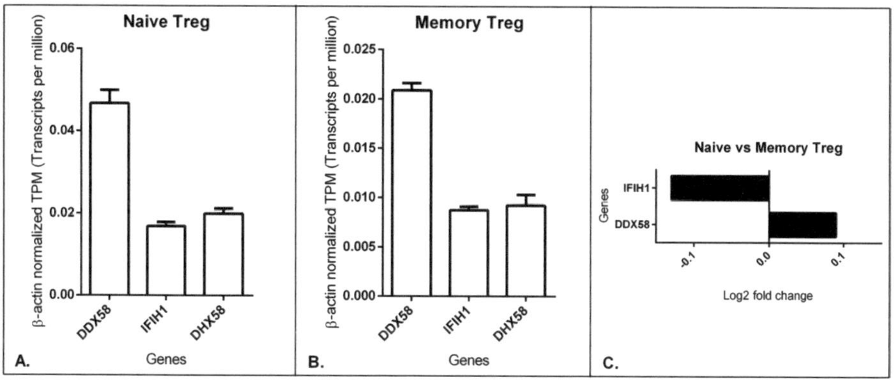

Figure 4.
RIG-I expression profile in naive Tregs and memory Tregs: Expression of RLRs in A: naive (CD3+, CD4+, CD25 high, CD45RA+, CD127low); B: memory (CD3+, CD4+, CD25 high, CD45RA-, CD127low) regulatory T cells. C: Expression of genes reported as TPM was normalized to β-actin (TPMgene/TPMβ-actin) for individual representation of naive and memory Tregs. Results are depicted as bar plots in GraphPad Prism.

	Biotype	Naive treg mean expression (TPM)	Memory treg mean expression (TPM)	Log 2 fold change	Adjusted p-value
DDX58	Protein coding	65.92	58.75	0.09	0.026
IFIH1	Protein coding	29.18	29.88	−0.13	0.0012

Table 4.
Expression levels of mRNA for each RIG-I like Receptor family member naïve and memory Tregs, p-values (https://dice-database.org/).

differentiation in a IFN regulatory transcription factor (IRF)-3 dependent manner [56]. Another study investigated the role of RNA-unprimed RIG-I (apo RIG-I) in Treg differentiation. In contrast to the suppression of Treg differentiation by RNA ligand primed RIG-I, this study showed that apo-RIG-I maintains the Treg/Th17 cell balance [57]. Taken together, ever expanding novel findings from large datasets and state-of-the-art approaches may change the way we evaluate PRRs of Tregs in most bewildering ways.

3. Functionality of regulatory T cells in disease in the context of PRRs

3.1 Regulatory T cells in infections

Immune system as a whole represents a quite complex and interacting vast network of cells and biochemical signals circulating in blood and tissues. Therefore, this complexity necessitates a tight regulation. Tregs maintain the homeostasis by suppressing the immune response after the infection is resolved. As discussed earlier, Tregs can be activated by a variety of pathogens and their suppressive functions may differ depending on the pathogen and the progression of the infection. Not only pathogens, but also non-pathogenic environments with endogenous proteins are essential in regulating Treg responses. Heat shock protein gp96 is a chaperone for several TLRs including TLR4 and acts as a ligand as well [58, 59]. Tregs suppress the T cell proliferation and cytokine release to protect the host against excessive immune response. There are few aspects still being investigated, especially which receptors are expressed in Tregs and regulate Tregs during viral, bacterial, and fungal infections [60]. In this part of the chapter, we will continue to discuss the Tregs in terms of infectious and non-infectious disease conditions.

3.1.1 Regulatory T cells in bacterial infections

Helicobacter *pylori* is a Gram-negative bacterium that infects the stomach and is a highly contagious pathogen [61]. Currently, studies covering PRRs in the context of Tregs are rather sparse when it comes to bacterial infections. However, TLR4 inhibition following H. *pylori* has been shown to increase the Foxp3 expression and the $CD4^{+}CD25^{+}Foxp3^{+}$ Treg cell numbers in the gastric mucosa and enhance the bacterial colonization. However, blocking Tregs led to the limited colonization of *H. pylori,* resulting in the reduced inflammatory responses. This study indicated that the crosstalk between TLR4 signaling pathway and Tregs is crucial for limiting *H. pylori* colonization and suppressing the inflammation of infected mice [62].

3.1.2 Regulatory T cells in fungus infections

Paracoccidioidomycosis (PCM) is an endemic disease caused by the fungus Paracoccidioides brasiliensis [63]. In PCM disease, regulation of Treg functions is mediated by PRRs such as TLRs, CLRs, and NLRs and downstream proteins like MyD88 [41, 64–67]. In one study, TLR2-deficient mice had a reduced number of Tregs along with an excessive immune response, suggesting that TLR2 is required for Treg expansion to control the inflammatory response [67]. The study utilized MyD88-deficient mice to further analyze the effect of the downstream effectors of TLR2 signaling pathway in Tregs. The absence of MyD88 resulted in the impaired T cell responses and uncontrolled spread of fungal infection in the murine model of PCM infection [65]. Another study showed that Treg proliferation is decreased in WT mice compared to TLR4 deficient mice, therefore, the level of infiltration of activated T cells and macrophages into the lung increased, resulting in severe infection [64]. Dectin-1 is a CTL receptor involved in the antifungal immune response [68]. In Dectin-1-deficient mice with P. *brasiliensis*, low levels of activated effector/memory CD4+ and CD8+ T cells along with increased CD4 + CD25 + Foxp3+ Treg levels let the infection spread to tissues, eventually controlling the severity of the disease and causing the increased mortality in mice [66]. Besides the membrane bound Dectin-1 receptor, NLRP3 inflammasome complex mediated the activation and secretion of IL1β and IL18 have been reported in P. *brasiliensis* infected mice. NLRP3 deficient mice infected with P. *brasiliensis* had an increased Th1/Th17 immune response and reduced Treg response [41].

During candidiasis, a fungal infection mediated by Candida *albicans,* TLR2 is of great importance to control Treg survival. In this study, Netea *et al.* reported that TLR2-deficient mice, but not TLR4, had increased Th1 immune response and a reduced IL10 producing CD4 + CD25+ Treg population as compared to the WT mice infected with C. *albicans*. Tregs were further stimulated with a TLR2 ligand, peptidoglycan, to assess the significance of TLR2 signaling. Findings from this study highlights the importance of TLR2 signaling in the maintenance of survival, expansion and suppression capacity of Treg and IL10 production in candidiasis [69].

3.1.3 Regulatory T cells in viral infections

One of the immune system mechanisms that are used to protect the host from viral infections is the recognition of viral nucleic acids by PRRs such as TLRs and RLRs [70]. The human genome encodes 10 different TLRs, four of which are responsible for the recognition of viral genome, and these are TLR3, TLR7, TLR8, and TLR9. Interestingly, unlike other TLRs located on the outer cell membrane, they are located in the endosomal membranes and induce downstream molecules through adaptor proteins [4]. RLRs sense viral RNAs in the cytosol. Among RLRs, RIG-I, and MDA5 have RNA helicase activity which give them the ability to bind viral RNA and induce immune response [46]. It is well established in the literature that innate immune cells are activated through these PRRs as the first line of host defense against viral attacks. Activated innate immune cells then phagocytose and process the virus to present it to naive T cells in the draining lymph nodes. Primed T cells eventually differentiate into different types of helper T cells including Tregs [60].

TLR2 and TLR4 have been the focus of numerous studies which emphasized the effects on regulatory T cells in the course of viral infections. During hepatitis C virus infection, Tregs were suggested to suppress the HCV-specific antiviral responses resulting in viral persistence [71, 72]. In a different study by Zhai *et al.*, the core protein of hepatitis C virus (HCVc) in the blood of HCV-infected patients has been shown to induce the proliferation of Tregs which subsequently hampered the CD4+ T cell proliferation and IFN-γ production. Since HCVc binds TLR2 on Kupffer cells and dendritic cells, just like the TLR2 agonist lipoteichoic acid, which leading to the similar processes, they proposed that HCVc mediated Treg expansion was TLR2 dependent [73]. In a separate study, the number of Tregs, TLR2 and TLR4 expression levels in the peripheral blood monocytes of chronic hepatitis C patients have been shown to elevate in parallel to viral load [74]. To this end, literature on TLRs seem to report consistent results with respect to Treg regulation.

Additionally, cytosolic RLRs have roles during viral infections. Amphiregulin, known as EGFR ligand, is produced mainly by Tregs in lungs during influenza A virus infection and it is important for tissue protection [75–77]. Interestingly, EGFR signaling has been suggested to suppress RIG-I signaling during viral infections [78, 79]. Thus, amphiregulin produced by regulatory T cells may reduce RIG-I signaling to increase survival during viral infections.

3.2 Regulatory T cells in autoimmune diseases

Autoimmunity can be defined as immunologic aberrations which exclusively exhibit abnormal self-antigen tolerance. PRRs can govern autoimmunity by playing pivotal roles in distinct immunological mechanisms [80, 81]. Autoimmune diseases have been associated with viral, bacterial and, more recently, fungal infections after detection by PRRs because of the reduced number of Treg cells and increased pro-inflammatory cytokines, such as IL17, IL22 and IL23, which drive the differentiation into CD4 Th17 T cells [82].

As we discussed previously in this chapter (**Figure 1**), TLRs are expressed and have functions in adaptive immune cells such as TCR alpha beta cells, TCR gamma beta T cells and regulatory T cells [83]. LPS induced TLR4 in $CD4^{+}CD25^{+}$ T cells have been shown to lead to activation and proliferation of Treg cells [12]. Although controversial, other TLRs including TLR5, TLR7, and TLR8 have been shown to express in human and murine $CD4^{+}CD45^{+}$ Tregs [11]. With regards to autoimmunity, using a cohort of MS patients who were helminth-infected or non-infected, Correale and Farez investigated the roles of retinoic acid (RA) and TLR2 in parasite mediated protection in MS patients. Helminth-activated DCs not only inhibited IL-17 and IFN-γ production via autoreactive T cells but also led to the immunoprotection which was attributed to the involvement of TLR2 and RA and the augmentation of $CD4^{+}CD25^{+}FOXP3^{+}$ Treg cells [84].

Multiple sclerosis (MS) is a central nervous system autoimmune disease [85] which is characterized by demyelination [86]. When healthy individuals were compared to MS patients, it was found that Tregs of the healthy group displayed higher TLR2 expression. Furthermore, the PBMC samples from these two separate groups were stimulated with an agonist of TLR1/2, Pam3Cys, which lowered Treg functions and induced Th17 in MS groups samples [87]. Another example of autoimmune disease is type 1 diabetes mellitus (T1DM) that is associated with pancreatic β cell deficiency which results in abnormal sugar level [88]. High mobility-group box (HMGB) proteins have a role to induce the innate immunity by interacting with nucleic acids and recruiting them to PRRs and they engage receptors for advanced glycation end products (RAGE) [89]. Wild *et al.* showed that HMGB1 enhanced IL-10 levels and prolonged survival of Treg cells [90]. Furthermore, the inhibition of HMGB1 during beta cell mass turnover at an early stage in NOD mice is followed by reduced incidence of diabetes. Additionally, TLR4 and RAGE were shown to be predominant HMGB1 receptors in Treg cells and blockade of either one diminished Treg instability whilst stimulation with recombinant HMGB1 remarkably increased the amount of phosphorylated downstream targets including PI3K, Akt and mTOR in Tregs [91]. Overall, PRRs and Tregs axis in certain immune conditions has been pending to be investigated more elaborately.

3.3 Regulatory T cells in asthma and allergy

Persistent inflammation with the excessive production of cytokines by the immune cells can be harmful which is associated with numerous diseases including asthma and allergy [92]. Asthma is an inflammatory disease of airways which is linked to excessive T helper cell type-2 (Th2) immunity. Both allergic and non-allergic stimuli including house dust mites (HDM), pollens, viral infections and tobacco smoke trigger a cascade of events resulting in chronic airway inflammation which then leads to the airway hyperresponsiveness (AHR) [93]. Th2 cells in the airway release specific cytokines including IL4, IL5, IL9, and IL13; thereby promoting eosinophilic inflammation and immunoglobulin E (IgE) production which in turn, triggers the release of other inflammatory mediators, such as leukotrienes and histamines [94]. One of the hallmarks of asthma pathogenesis is the enhanced Th2 response and the inadequate differentiation and functional defects of Tregs. Baatjes *et al.* has reported that $CD4^{+}CD25^{high}Foxp3^{+}$ Tregs were lower in the peripheral blood of the asthma patients than non-asthmatic individuals [95]. *In vivo* animal model studies have shown that IL10 and TGFβ secreted by Treg remarkably suppressed the airway inflammation and AHR [96] while blocking IL10 and TGFβ worsened the airway inflammation and AHR [97].

Several studies also revealed the involvement of PRRs, especially TLRs and NLRs in asthma susceptibility. Simpson *et al.* firstly discovered the upregulation of TLR2, TLR4, and pro-inflammatory cytokines, IL1β and IL18, in neutrophilic asthma [98]. Another study reported that TLR2 activation induces Treg and long-term suppression of asthma symptoms in OVA-sensitized mice [99]. Moreover, the important roles of

TLR7 in alleviation of airway inflammation, promoting Th1 immune responses and reversing AHR have been shown [100]. Meng *et al.* reported that TLR7 stimulation suppressed eosinophilic inflammation by reducing Th2 cytokines IL4 and IL5, eotaxin and IgE in numerous animal models of asthma [101]. Yet another study added that treatment with TLR7 agonist R848 induced Treg cell-mediated suppression of asthma symptoms in OVA-sensitized and challenged mice [102].

Although limited, the involvement of inflammasome activation in asthmatic airway inflammations has been studied as well. The prolonged administration of IL1β, an inflammasome dependent cytokine, has been shown to induce AHR [103]. Also, increased levels of IL1β in the serum and BALF of asthmatic patients were decreased after glucocorticoids inhalation [104]. Significantly higher inflammasome depent IL18 levels in the serum of asthma patients were detected [105]. Moreover, Simpson *et al.* reported the elevated expression of the NLRP3 inflammasome in patients with neutrophilic asthma [106]. Another recent study using HDM-induced mouse models of allergic airway inflammation reported elevated expression of NLRP3, NLRC4, NLRC5, and caspase-1 genes as well as pro- IL1β levels in the lungs, while mature IL1β was not observed which suggested that inflammasome components are upregulated even if they do not form functional inflammasome complexes [107]. Unfortunately, studies as to the effects of inflammasomes on Treg functions in allergy and asthma conditions are limited. One study demonstrated that intranasal stimulation with NOD2 ligand disrupted the generation of Tregs and subsequently induced the development of eosinophil-associated airway inflammation [108]. Altogether, targeting NLRs and inflammasome components can be another potential therapeutic approach for the control of airway inflammation.

3.4 Regulatory T cells in cancer

Countless pathological conditions involve infections and tissue damage leading to chronic inflammation after the activation of PRRs. Innate immunity and PRRs in cancer initiation and progression are extensively studied because PRRs are expressed in different tumor tissues, such as lung, breast, colon, gastric cancer, and melanoma [21, 109]. The PRR activation in cancer cells can stimulate the production of many cytokines, chemokines, hormones, and vascular-promoting factors to induce the formation of an inflammatory tumor microenvironment that promotes the tumor progression [110]. The activation of PRRs on antigen presenting immune cells can induce dendritic cells, tumor-associated macrophages, and B cells for the generation of tumor-specific T cell responses. Tregs are found in tumor microenvironment and are able to suppress anti-tumor immune responses which is required for escaping immune system thereby cancer progression.

Signaling through PRRs results in robust pro-inflammatory responses by promoting antigen presenting cells and orchestrating adaptive immunity against tumor associated antigens [110]. Indeed, PRR ligands can both stimulate tumors and tumor-infiltrating immune cells to secrete cytokines and chemokines which modulate immune cell polarization and reprogramming the tumor microenvironment to reinforce innate and adaptive anti-tumor immunity [111]. Even though the roles of NLRs and RLRs in tumor immunity still largely unknown, TLRs have significant roles in stimulating DC maturation, antigen uptake and presentation, and the differentiation of $CD4^+$ T cells. Additionally, Nyirenda *et al.* have reported the reversion of immunosuppressive skills of Tregs upon TLR stimulation and this is especially interesting for cancer research due to their activity in tumor microenvironment [112]. Given what we know about TLRs and their ligands currently, different TLR agonists have been used in anticancer therapies. Studies on TLR8 have demonstrated that adoptive transfer of TLR8 ligand-stimulated Treg cells reduced the tumor growth in mice [14] by reprogramming Treg glucose metabolism [113]. On the contrary, peritumoral administration of TLR5 ligand flagellin did not affect the growth of murine breast carcinoma D2F2 [114].

Several inflammasome forming NLRs including NLRP1, NLRP3, NLRP6, and NLRC4 may both have protective and detrimental roles in tumor development by their modulation of innate and adaptive immunity, apoptosis and differentiation [115]. On one hand, Janowski *et al.* has reported the protective role of NLRC4 in melanoma progression independent of the inflammasome components ASC and caspase-1 [116]. On the other hand, a recent study has shown that in metastatic melanoma and sarcoma models, NLRP3 inflammasome activation increased Treg population while inhibiting both NK and T-cell mediated anti-tumor immunity [117]. Additionally, recent studies demonstrated that NLRP3 inflammasome inhibitor (MCC950) reduced IL1β production and Tregs in head and neck squamous cell carcinoma mouse model [118]; and NLRP3 inhibitor (OLT1177) reduced melanoma growth and $Foxp3^+$ cells in tumor microenvironment, and when given in combination with anti-PD-1 therapy, its efficacy increased as compared to monotherapy [119].

Even though CLRs are expressed by dendritic cells, they trigger distinct signaling pathways which induce the expression of cytokines and ultimately determine the T cell differentiation. There are several CLR agonists or antagonists that can be used as anti-cancer drugs, such as β-glucan as dectin-1 agonists [120]. With this, Osorio *et al.* have shown that a bacterial β-glucan, curdlan, skewed Tregs toward Th17 cells both in *vitro* and in *vivo* using 4 T1 mouse mammary tumor models [121]. Also, another study has demonstrated that LSECtin, a type II transmembrane protein, which belongs to the C-type lectin receptor superfamily inhibited the proliferation of tumor-specific effector T cells and induce more IL10 production from Treg cells [122].

In addition to these, cancer cells may mimic viral infections to activate interferon response pathway, and activation of RLR signaling in cancer cells may trigger cell death, activation of innate immune cells in tumor microenvironment or increased recruitment of adaptive immune cells into poorly immunogenic tumors [123]. RLRs could inhibit growth or induce apoptosis of different types of cancer cells upon recognition of RNA ligands. Jiang *et al.* has noted that intratumoral delivery of SLR14, RIG-I agonist induced strong anti-tumor immune responses through the reduction of CD4 + FoxP3+ Treg cells and induction of $CD8^+$ T lymphocytes and NK cells [124]. In another study, high RIG-I expression in ovarian cancer was associated with increased FoxP3 expression and enriched PD-L1 and PD-1 mRNA expression [125]. Taken together, it is noteworthy that even though cancer disease and tumor microenvironment are highly heterogeneous, findings from completed and ongoing research raise the possibility of new targets for the treatment of cancer.

4. Conclusions

To study infectious and immune system related diseases is specifically difficult due to the genetic diversity of hosts and pathogens, the ever-changing nature of infection as it progresses, and the secession of host responses during the course of infection and the disease progression. Despite these challenges, utilization of more sophisticated, contemporary immunogenetic methods and tools such as single cell sequencing, high throughput screenings, computational modeling along with the availability of novel *in vivo* disease models, 3D organoid cultures perhaps lead to exciting outcomes for the long-term control of both infectious diseases and non-infectious immune system related diseases. These approaches will eventually lay the foundation of a framework to understand the interactive relationship between PRRs and regulatory T cells.

Acknowledgements

Authors thank ITU Scientific Research Projects Department Project No 43336 and 40713.

Conflict of interest

The authors declare no conflict of interest.

Author details

Ece Oylumlu, Goksu Uzel, Lubeyne Durmus, Meric Tas, Damla Gunes and Ceren Ciraci*
College of Science and Letters, Molecular Biology and Genetics Department, Istanbul Technical University, Istanbul, Turkey

*Address all correspondence to: ciracic@itu.edu.tr

References

[1] Kondělková K, Vokurková D, Krejsek J, Borská L, Fiala Z, Ctirad A. Regulatory T Cells (TREG) and their Roles in Immune System with Respect to Immunopathological Disorders. Facultas Medica Hradec Králové: Acta medica (Hradec Králové)/Universitas Carolina; 2010. DOI: 10.14712/18059694.2016.63

[2] Mills KHG. Regulatory T cells: Friend or foe in immunity to infection? Nature Reviews Immunology. 2004;**4**: 841-855. DOI: 10.1038/nri1485

[3] Belkaid Y, Rouse BT. Natural regulatory T cells in infectious disease. Nature Immunology. 2005;**6**:353-360. DOI: 10.1038/ni1181

[4] Kawasaki T, Kawai T. Toll-like receptor signaling pathways. Frontiers in Immunology. 2014;5:1-8. DOI: 10.3389/fimmu.2014.00461

[5] Dai J, Liu B, Li Z. Regulatory T cells and Toll-like receptors: What is the missing link? International Immunopharmacology. 2009;**9**:528-533. DOI: 10.1016/j.intimp.2009.01.027

[6] Bell JK, Mullen GED, Leifer CA, Mazzoni A, Davies DR, Segal DM. Leucine-rich repeats and pathogen recognition in Toll-like receptors. Trends in Immunology. 2003;**24**:528-533. DOI: 10.1016/S1471-4906(03)00242-4

[7] Botos I, Segal DM, Davies DR. The structural biology of Toll-like receptors. Structure. 2011;**19**:447-459. DOI: 10.1016/j.str.2011.02.004

[8] Sameer AS, Nissar S. Toll-Like Receptors (TLRs): Structure, functions, signaling, and role of their polymorphisms in colorectal cancer susceptibility. BioMed Research International. 2021;**2021**:1-14. DOI: 10.1155/2021/1157023

[9] Sutmuller RPM, Garritsen A, Adema GJ. Regulatory T cells and toll-like receptors: Regulating the regulators. Annals of the Rheumatic Diseases. 2007;**66**(Suppl 3):iii91-iii95. DOI: 10.1136/ard.2007.078535

[10] Li MO, Rudensky AY. T cell receptor signalling in the control of regulatory T cell differentiation and function. Nature Reviews Immunology. 2016;**16**:220-233. DOI: 10.1038/nri.2016.26

[11] Van Maren WWC, Jacobs JFM, De Vries IJM, Nierkens S, Adema GJ. Toll-like receptor signalling on tregs: To suppress or not to suppress? Immunology. 2008;**124**:445-452. DOI: 10.1111/j.1365-2567.2008.02871.x

[12] Caramalho I, Lopes-Carvalho T, Ostler D, Zelenay S, Haury M, Demengeot J. Regulatory T cells selectively express toll-like receptors and are activated by lipopolysaccharide. The Journal of Experimental Medicine. 2003;**197**:403-411. DOI: 10.1084/jem.20021633

[13] Crellin NK, Garcia RV, Hadisfar O, Allan SE, Steiner TS, Levings MK. Human CD4 + T cells express TLR5 and its ligand flagellin enhances the suppressive capacity and expression of FOXP3 in CD4 + CD25 + T regulatory cells. Journal of Immunology. 2005:175. DOI: 10.4049/jimmunol.175.12.8051

[14] Peng G, Guo Z, Kiniwa Y, Voo KS, Peng W, Fu T, et al. Immunology: Toll-like receptor 8-mediated reversal of CD4+ regulatory T cell function. Science (80-). 2005:309. DOI: 10.1126/science.1113401

[15] Chiffoleau E, Heslan JM, Heslan M, Louvet C, Condamine T, Cuturi MC. TLR9 ligand enhances proliferation of rat CD4+ T cell and modulates suppressive activity mediated by CD4+ CD25+ T cell. International Immunology. 2007;**19**: 193-201. DOI: 10.1093/intimm/dxl136

[16] Schmiedel BJ, Singh D, Madrigal A, Valdovino-Gonzalez AG, White BM, Zapardiel-Gonzalo J, et al. Impact of genetic polymorphisms on human immune cell gene expression. Cell. 2018;**175**:1701-1715. DOI: 10.1016/j.cell.2018.10.022

[17] Monaco G, Lee B, Xu W, Mustafah S, Hwang YY, Carré C, et al. RNA-Seq signatures normalized by mRNA abundance allow absolute deconvolution of human immune cell types. Cell Reports. 2019;**26**:1627-1640. DOI: 10.1016/j.celrep.2019.01.041

[18] Ota M, Nagafuchi Y, Hatano H, Ishigaki K, Terao C, Takeshima Y, et al. Dynamic landscape of immune cell-specific gene regulation in immune-mediated diseases. Cell. 2021;**184**: 3006-3021. DOI: 10.1016/j.cell.2021.03.056

[19] Geijtenbeek TBH, Gringhuis SI. Signalling through C-type lectin receptors: Shaping immune responses. Nature Reviews Immunology. 2009;**9**:465-479. DOI: 10.1038/nri2569

[20] Lepenies B, Lee J, Sonkaria S. Targeting C-type lectin receptors with multivalent carbohydrate ligands. Advanced Drug Delivery Reviews. 2013;**65**:1271-1281. DOI: 10.1016/j.addr.2013.05.007

[21] Li D, Wu M. Pattern recognition receptors in health and diseases. Signal Transduction and Targeted Therapy. 2021;**6**:1-24. DOI: 10.1038/s41392-021-00687-0

[22] Rieber N, Gazendam RP, Freeman AF, Hsu AP, Collar AL, Sugui JA, et al. Extrapulmonary Aspergillus infection in patients with CARD9 deficiency. JCI Insight. 2021;**1**:1-13. DOI: 10.1172/jci.insight.89890

[23] Speakman EA, Dambuza IM, Salazar F, Brown GD. T cell antifungal immunity and the role of C-type lectin receptors. Trends in Immunology. 2020;**41**:61-76. DOI: 10.1016/j.it.2019.11.007

[24] Singh SK, Streng-Ouwehand I, Litjens M, Weelij DR, García-Vallejo JJ, van Vliet SJ, et al. Characterization of murine MGL1 and MGL2 C-type lectins: Distinct glycan specificities and tumor binding properties. Molecular Immunology. 2009;**46**:1240-1249. DOI: 10.1016/j.molimm.2008.11.021

[25] Zizzari IG, Napoletano C, Battisti F, Rahimi H, Caponnetto S, Pierelli L, et al. MGL receptor and immunity: When the ligand can make the difference. Journal of Immunology Research. 2015;**2015**:1-8. DOI: 10.1155/2015/450695

[26] van Vliet SJ, van Liempt E, Saeland E, Aarnoudse CA, Appelmelk B, Irimura T, et al. Carbohydrate profiling reveals a distinctive role for the C-type lectin MGL in the recognition of helminth parasites and tumor antigens by dendritic cells. International Immunology. 2005;**17**:261-669. DOI: 10.1093/intimm/dxh246

[27] Raes G, Brys L, Dahal BK, Brandt J, Grooten J, Brombacher F, et al. Macrophage galactose-type C-type lectins as novel markers for alternatively activated macrophages elicited by parasitic infections and allergic airway inflammation. Journal of Leukocyte Biology. 2005;**77**:321-327. DOI: 10.1189/jlb.0304212

[28] Napoletano C, Zizzari IG, Rughetti A, Rahimi H, Irimura T, Clausen H, et al. Targeting of macrophage galactose-type C-type lectin (MGL) induces DC signaling Regulatory T Cells and activation. European Journal of Immunology. 2012;**42**:936-945. DOI: 10.1002/eji.201142086

[29] Saijo S, Iwakura Y. Dectin-1 and Dectin-2 in innate immunity against fungi. International Immunology. 2011;**23**:467-472. DOI: 10.1093/intimm/dxr046

[30] Wevers BA, Kaptein TM, Zijlstra-Willems EM, Theelen B, Boekhout T, Geijtenbeek TBH, et al. Fungal engagement of the C-type lectin mincle suppresses dectin-1-induced antifungal immunity. Cell Host & Microbe. 2014;**15**:494-505. DOI: 10.1016/j.chom.2014.03.008

[31] LeibundGut-Landmann S, Groß O, Robinson MJ, Osorio F, Slack EC, Tsoni SV, et al. Syk- and CARD9-dependent coupling of innate immunity to the induction of T helper cells that produce interleukin 17. Nature Immunology. 2007;**8**:630-638. DOI: 10.1038/ni1460

[32] Sánchez-Díaz R, Blanco-Dominguez R, Lasarte S, Tsilingiri K, Martín-Gayo E, Linillos-Pradillo B, et al. Thymus-derived regulatory T cell development is regulated by C-type lectin-mediated BIC/MicroRNA 155 expression. Molecular and Cellular Biology. 2017;**37**: 1-18. DOI: 10.1128/mcb.00341-16

[33] Motta V, Soares F, Sun T, Philpott DJ. Nod-like receptors: Versatile cytosolic sentinels. Physiological Reviews. 2015;**95**: 149-178. DOI: 10.1152/physrev.00009.2014

[34] Werts C, Rubino S, Ling A, Girardin SE, Philpott DJ. Nod-like receptors in intestinal homeostasis, inflammation, and cancer. Journal of Leukocyte Biology. 2011;**90**:9594-9627. DOI: 10.1189/jlb.0411183

[35] Lamkanfi M, Dixit VM. Mechanisms and functions of inflammasomes. Cell. 2014;**157**:1013-1022. DOI: 10.1016/j.cell.2014.04.007

[36] Davis BK, Wen H, Ting JP-Y. The inflammasome NLRs in immunity, inflammation, and associated diseases. Annual Review of Immunology. 2011;**29**:707-735. DOI: 10.1146/annurev-immunol-031210-101405

[37] Pellegrini C, Antonioli L, Lopez-Castejon G, Blandizzi C, Fornai M. Canonical and non-canonical activation of NLRP3 inflammasome at the crossroad between immune tolerance and intestinal inflammation. Frontiers in Immunology. 2017;**8**:1-12. DOI: 10.3389/fimmu.2017.00036

[38] Ciraci C, Janczy JR, Jain N, Haasken S, Silva CP, Benjamim CF, et al. Immune complexes indirectly suppress the generation of Th17 responses In Vivo. PLoS One. 2016;**11**:1-15. DOI: 10.1371/journal.pone.0151252

[39] Liu D, Rhebergen AM, Eisenbarth SC. Licensing adaptive immunity by NOD-like receptors. Frontiers in Immunology. 2013;**4**:1-17. DOI: 10.3389/fimmu.2013.00486

[40] Park SH, Ham S, Lee A, Möller A, Kim TS. NLRP3 negatively regulates Treg differentiation through Kpna2-mediated nuclear translocation. The Journal of Biological Chemistry. 2019;**294**:17951-17961. DOI: 10.1074/jbc.RA119.010545

[41] Feriotti C, de Araújo EF, Loures FV, Costa TA da, Galdino NA de L, Zamboni DS, et al. NOD-like receptor P3 inflammasome controls protective Th1/Th17 immunity against pulmonary

paracoccidioidomycosis. Frontiers in Immunology 2017;**8**:1-15. DOI:10.3389/fimmu.2017.00786

[42] Rahman MK, Midtling EH, Svingen PA, Xiong Y, Bell MP, Tung J, et al. The pathogen recognition receptor NOD2 regulates human FOXP3 + T cell survival. Journal of Immunology. 2010;**184**:7247-7256. DOI: 10.4049/jimmunol.0901479

[43] Egarnes B, Gosselin J. Contribution of regulatory T cells in nucleotide-binding oligomerization domain 2 response to influenza virus infection. Frontiers in Immunology. 2018;**9**:1-12. DOI: 10.3389/fimmu.2018.00132

[44] Loo YM, Gale M. Immune signaling by RIG-I-like receptors. Immunity. 2011;**34**:680-692. DOI: 10.1016/j.immuni.2011.05.003

[45] McNab F, Mayer-Barber K, Sher A, Wack A, O'Garra A. Type I interferons in infectious disease. Nature Reviews Immunology. 2015;**15**:87-103. DOI: 10.1038/nri3787

[46] Rehwinkel J, Gack MU. RIG-I-like receptors: Their regulation and roles in RNA sensing. Nature Reviews Immunology. 2020;**20**:537-551. DOI: 10.1038/s41577-020-0288-3

[47] Zou J, Chang M, Nie P, Secombes CJ. Origin and evolution of the RIG-I like RNA helicase gene family. BMC Evolutionary Biology. 2009;**9**:1-14. DOI: 10.1186/1471-2148-9-85

[48] Childs KS, Randall RE, Goodbourn S. LGP2 plays a critical role in sensitizing mda-5 to activation by double-stranded RNA. PLoS One. 2013;**8**:1-8. DOI: 10.1371/journal.pone.0064202

[49] Rawling DC, Pyle AM. Parts, assembly and operation of the RIG-I family of motors. Current Opinion in Structural Biology. 2014;**25**:25-33. DOI: 10.1016/j.sbi.2013.11.011

[50] Sellati TJ, Sahay B. Cells of innate immunity: Mechanisms of activation. Pathobiology of Human Disease: A Dynamic Encyclopedia of Disease Mechanisms. 2014:258-274. DOI: 10.1016/B978-0-12-386456-7.01804-9

[51] Kubo T, Hatton RD, Oliver J, Liu X, Elson CO, Weaver CT. Regulatory T cell suppression and anergy are differentially regulated by proinflammatory cytokines produced by TLR-activated dendritic cells. Journal of Immunology. 2004;**173**:7249-7258. DOI: 10.4049/jimmunol.173.12.7249

[52] Sutmuller RPM, Den Brok MHMGM, Kramer M, Bennink EJ, Toonen LWJ, Kullberg BJ, et al. Toll-like receptor 2 controls expansion and function of regulatory T cells. The Journal of Clinical Investigation. 2006;**116**:485-494. DOI: 10.1172/JCI25439

[53] Anz D, Koelzer VH, Moder S, Thaler R, Schwerd T, Lahl K, et al. Immunostimulatory RNA Blocks Suppression by regulatory T Cells. Journal of Immunology. 2010;**184**: 939-946. DOI: 10.4049/jimmunol.0901245

[54] Kawai T, Takahashi K, Sato S, Coban C, Kumar H, Kato H, et al. IPS-1, an adaptor triggering RIG-I- and Mda5-mediated type I interferon induction. Nature Immunology. 2005;**6**:981-988. DOI: 10.1038/ni1243

[55] Suthar MS, Ma DY, Thomas S, Lund JM, Zhang N, Daffis S, et al. IPS-1 is essential for the control of west nile virus infection and immunity. PLoS Pathogens. 2010;**6**:1-15. DOI: 10.1371/journal.ppat.1000757

[56] Xu P, Bailey-Bucktrout S, Xi Y, Xu D, Du D, Zhang Q, et al. Innate antiviral host defense attenuates TGF-β function through IRF3-mediated suppression of smad signaling. Molecular Cell. 2014;**56**: 723-737. DOI: 10.1016/j.molcel.2014.11.027

[57] Yang H, Guo H-Z, Li X-Y, Lin J, Zhang W, Zhao J-M, et al. Viral RNA–unprimed Rig-I restrains Stat3 activation in the modulation of regulatory T Cell/Th17 cell balance. Journal of Immunology. 2017;**199**:119-128. DOI: 10.4049/jimmunol.1700366

[58] Vabulas RM, Wagner H, Schild H. Heat shock proteins as ligands of Toll-like receptors. Current Topics in Microbiology and Immunology. 2002;**270**:169-184. DOI: 10.1007/978-3-642-59430-4_11

[59] Dai J, Liu B, Ngoi SM, Sun S, Vella AT, Li Z. TLR4 hyperresponsiveness via cell surface expression of heat shock protein gp96 potentiates suppressive function of regulatory T cells. Journal of Immunology. 2007;**178**:3219-3225. DOI: 10.4049/jimmunol.178.5.3219

[60] Boer MC, Joosten SA, Ottenhoff THM. Regulatory T-cells at the interface between human host and pathogens in infectious diseases and vaccination. Frontiers in Immunology. 2015;**6**:1-15. DOI: 10.3389/fimmu.2015.00217

[61] Hooi JKY, Lai WY, Ng WK, Suen MMY, Underwood FE, Tanyingoh D, et al. Global prevalence of helicobacter pylori infection: Systematic review and meta-analysis. Gastroenterology. 2017;**153**:420-429. DOI: 10.1053/j.gastro.2017.04.022

[62] Gong Y, Tao L, Jing L, Liu D, Hu S, Liu W, et al. Association of TLR4 and treg in Helicobacter pylori colonization and inflammation in mice. PLoS One. 2016;**11**:1-16. DOI: 10.1371/journal.pone.0149629

[63] Marques-da-Silva SH, Rodrigues AM, De Hoog GS, Silveira-Gomes F, De Camargo ZP. Case report: Occurrence of paracoccidioides lutzii in the amazon region: Description of two cases. The American Journal of Tropical Medicine and Hygiene. 2012;**87**:710-714. DOI: 10.4269/ajtmh.2012.12-0340

[64] Loures FV, Pina A, Felonato M, Araújo EF, Leite KRM, Calich VLG. Toll-like receptor 4 signaling leads to severe fungal infection associated with enhanced proinflammatory immunity and impaired expansion of regulatory T cells. Infection and Immunity. 2010:78. DOI: 10.1128/IAI.01198-09

[65] Loures FV, Pina A, Felonato M, Feriotti C, de Araújo EF, Calich VLG. MyD88 signaling is required for efficient innate and adaptive immune responses to Paracoccidioides brasiliensis infection. Infection and Immunity. 2011:79. DOI: 10.1128/IAI.00375-10

[66] Loures FV, Araújo EF, Feriotti C, Bazan SB, Costa TA, Brown GD, et al. Dectin-1 induces M1 macrophages and prominent expansion of CD8 +IL-17+ cells in pulmonary paracoccidioidomycosis. The Journal of Infectious Diseases. 2014:210. DOI: 10.1093/infdis/jiu136

[67] Loures FV, Pina A, Felonato M, Calich VLG. TLR2 is a negative regulator of Th17 cells and tissue pathology in a pulmonary model of fungal infection. Journal of Immunology. 2009:183. DOI: 10.4049/jimmunol.0801599

[68] Tang J, Lin G, Langdon WY, Tao L, Zhang J. Regulation of C-type lectin receptor-mediated antifungal immunity. Frontiers in Immunology. 2018;**9**:1-12. DOI: 10.3389/fimmu.2018.0012

[69] Netea MG, Sutmuller R, Hermann C, Van der Graaf CAA, Van der Meer JWM, van Krieken JH, et al. Toll-like receptor 2 suppresses immunity against candida albicans through induction of IL-10 and regulatory T cells. Journal of Immunology. 2004;**172**:3712-3718. DOI: 10.4049/jimmunol.172.6.3712

[70] Hotz C, Roetzer LC, Huber T, Sailer A, Oberson A, Treinies M, et al. TLR and RLR signaling are reprogrammed in opposite directions after detection of viral infection. Journal of Immunology. 2015;**195**:4387-4395. DOI: 10.4049/jimmunol.1500079

[71] Burton JR, Klarquist J, Im KA, Smyk-Pearson S, Golden-Mason L, Castelblanco N, et al. Prospective analysis of effector and regulatory CD4+ T cells in chronic HCV patients undergoing combination antiviral therapy. Journal of Hepatology. 2008;**49**: 329-338. DOI: 10.1016/j.jhep.2008.05.020

[72] Cabrera R, Tu Z, Xu Y, Firpi RJ, Rosen HR, Liu C, et al. An immunomodulatory role for CD4+CD25+ regulatory T lymphocytes in hepatitis C virus infection. Hepatology. 2004;**40**: 1062-1071. DOI: 10.1002/hep.20454

[73] Zhai N, Chi X, Li T, Song H, Li H, Jin X, et al. Hepatitis C virus core protein triggers expansion and activation of CD4+CD25+regulatory T cells in chronic hepatitis C patients. Cellular & Molecular Immunology. 2015;**12**:743-749. DOI: 10.1038/cmi.2014.119

[74] Wang JP, Zhang Y, Wei X, Li J, Nan XP, Yu HT, et al. Circulating toll-like receptor (TLR) 2, TLR4, and regulatory T cells in patients with chronic hepatitis C. APMIS. 2010;**118**:261-270. DOI: 10.1111/j.1600-0463.2010.02586.x

[75] Arpaia N, Green JA, Moltedo B, Arvey A, Hemmers S, Yuan S, et al. A distinct function of regulatory T cells in tissue protection. Cell. 2015;**162**:1078-1089. DOI: 10.1016/j.cell.2015.08.021

[76] Burzyn D, Kuswanto W, Kolodin D, Shadrach JL, Cerletti M, Jang Y, et al. A special population of regulatory T cells potentiates muscle repair. Cell. 2013;**155**:1282-1295. DOI: 10.1016/j.cell.2013.10.054

[77] Harb H, Benamar M, Lai PS, Contini P, Griffith JW, Crestani E, et al. Notch4 signaling limits regulatory T-cell-mediated tissue repair and promotes severe lung inflammation in viral infections. Immunity. 2021;**54**: 1186-1199. DOI: 10.1016/j.immuni.2021.04.002

[78] Oshiumi H, Miyashita M, Okamoto M, Morioka Y, Okabe M, Matsumoto M, et al. DDX60 is involved in RIG-I-dependent and independent antiviral responses, and its function is attenuated by virus-induced EGFR activation. Cell Reports. 2015;**11**:1193-1207. DOI: 10.1016/j.celrep.2015.04.047

[79] Ueki IF, Min-Oo G, Kalinowski A, Ballon-Landa E, Lanier LL, Nadel JA, et al. Respiratory virus-induced EGFR activation suppresses IRF1-dependent interferon λ and antiviral defense in airway epithelium. The Journal of Experimental Medicine. 2013;**210**:1929-1936. DOI: 10.1084/jem.20121401

[80] Chen L, Qi CS, Min LZ, Jun HS, Ping ZJ. NOD-like receptors in autoimmune diseases. Acta Pharmacologica Sinica. 2021. DOI: 10.1038/s41401-020-00603-2

[81] Ciraci C. Physiology and pathology of autoinflammation: NOD like receptors in autoinflammation and autoimmunity. Physiology and Pathology of Immunology. 2017. DOI: 10.5772/intechopen.70484

[82] Roe K. How major fungal infections can initiate severe autoimmune diseases. Microbial Pathogenesis. 2021;**161**:1-5. DOI: 10.1016/j.micpath.2021.105200

[83] Gibson J, Gow N, Wong SYC. Expression and functions of innate pattern recognition receptors in T and B cells. Immunology, Endocrine & Metabolic Agents in Medicinal Chemistry. 2012;**10**:11-20. DOI: 10.2174/187152210791171304

[84] Correale J, Farez MF. Parasite infections in multiple sclerosis modulate immune responses through a retinoic acid–dependent pathway. Journal of Immunology. 2013;**191**:3827-3837. DOI: 10.4049/jimmunol.1301110

[85] Castrop F, Haslinger B, Hemmer B, Buck D. Review of the pharmacoeconomics of early treatment of multiple sclerosis using interferon beta. Neuropsychiatric Disease and Treatment. 2013;**9**:1339-1349. DOI: 10.2147/NDT.S33949

[86] Podda G, Nyirenda M, Crooks J, Gran B. Innate immune responses in the CNS: Role of toll-like receptors, mechanisms, and therapeutic opportunities in multiple sclerosis. Journal of Neuroimmune Pharmacology. 2013. DOI: 10.1007/s11481-013-9483-3

[87] Nyirenda MH, Morandi E, Vinkemeier U, Constantin-Teodosiu D, Drinkwater S, Mee M, et al. TLR2 stimulation regulates the balance between regulatory T cell and Th17 function: A novel mechanism of reduced regulatory T cell function in multiple sclerosis. Journal of Immunology. 2015;**194**:5761-5774. DOI: 10.4049/jimmunol.1400472

[88] Katsarou A, Gudbjörnsdottir S, Rawshani A, Dabelea D, Bonifacio E, Anderson BJ, et al. Type 1 diabetes mellitus. Nature Reviews Disease Primers. 2017;**3**:1-17. DOI: 10.1038/nrdp.2017.16

[89] Sims GP, Rowe DC, Rietdijk ST, Herbst R, Coyle AJ. HMGB1 and RAGE in inflammation and cancer. Annual Review of Immunology. 2010;**28**:367-388. DOI: 10.1146/annurev.immunol.021908.132603

[90] Wild CA, Bergmann C, Fritz G, Schuler P, Hoffmann TK, Lotfi R, et al. HMGB1 conveys immunosuppressive characteristics on regulatory and conventional T cells. International Immunology. 2012;**24**:485-494. DOI: 10.1093/intimm/dxs051

[91] Zhang J, Chen L, Wang F, Zou Y, Li J, Luo J, et al. Extracellular HMGB1 exacerbates autoimmune progression and recurrence of type 1 diabetes by impairing regulatory T cell stability. Diabetologia. 2020;**63**:987-1001. DOI: 10.1007/s00125-020-05105-8

[92] Dong Z, Xiong L, Zhang W, Gibson PG, Wang T, Lu Y, et al. Holding the inflammatory system in check: TLRs and their targeted therapy in asthma. Mediators of Inflammation. 2016;**2**. DOI: 10.1155/2016/2180417

[93] Quirt J, Hildebrand KJ, Mazza J, Noya F, Kim H. Asthma. Allergy, Asthma and Clinical Immunology. 2018. DOI: 10.1186/s13223-018-0279-0

[94] Lemanske RF, Busse WW. Asthma: Clinical expression and molecular mechanisms. The Journal of Allergy and Clinical Immunology. 2010;**125**:S95-S102. DOI: 10.1016/j.jaci.2009.10.047

[95] Baatjes AJ, Smith SG, Watson R, Howie K, Murphy D, Larché M, et al. T regulatory cell phenotypes in peripheral blood and bronchoalveolar lavage from non-asthmatic and asthmatic subjects. Clinical and Experimental Allergy.

2015;**45**:1654-1662. DOI: 10.1111/cea.12594

[96] Fu CL, Chuang YH, Chau LY, Chiang BL. Effects of adenovirus-expressing IL-10 in alleviating airway inflammation in asthma. The Journal of Gene Medicine. 2006;**8**:1393-1399. DOI: 10.1002/jgm.974

[97] Oh JW, Seroogy CM, Meyer EH, Akbari O, Berry G, Fathman CG, et al. CD4 T-helper cells engineered to produce IL-10 prevent allergen-induced airway hyperreactivity and inflammation. The Journal of Allergy and Clinical Immunology. 2002;**110**:460-468. DOI: 10.1067/mai.2002.127512

[98] Simpson JL, Grissell TV, Douwes J, Scott RJ, Boyle MJ, Gibson PG. Innate immune activation in neutrophilic asthma and bronchiectasis. Thorax. 2007:62. DOI: 10.1136/thx.2006.061358

[99] Nawijn MC, Motta AC, Gras R, Shirinbak S, Maazi H, van Oosterhout AJM. TLR-2 activation induces regulatory T cells and long-term suppression of asthma manifestations in mice. PLoS One. 2013;**8**:1-9. DOI: 10.1371/journal.pone.0055307

[100] Duechs MJ, Hahn C, Benediktus E, Werner-Klein M, Braun A, Hoymann HG, et al. TLR agonist mediated suppression of allergic responses is associated with increased innate inflammation in the airways. Pulmonary Pharmacology & Therapeutics. 2011;**24**:203-214. DOI: 10.1016/j.pupt.2010.12.009

[101] Meng L, He X, Zhu W, Yang X, Jiang C, Sun Q, et al. TLR3 and TLR7 modulate IgE production in antigen induced pulmonary inflammation via influencing IL-4 expression in immune organs. PLoS One. 2011;**6**:1-11. DOI: 10.1371/journal.pone.0017252

[102] Pham Van L, Bardel E, Gregoire S, Vanoirbeek J, Schneider E, Dy M, et al. Treatment with the TLR7 agonist R848 induces regulatory T-cell-mediated suppression of established asthma symptoms. European Journal of Immunology. 2011;**41**:1992-1999. DOI: 10.1002/eji.201040914

[103] Zhang Y, Xu CB, Cardell LO. Long-term exposure to IL-1β enhances Toll-IL-1 receptor-mediated inflammatory signaling in murine airway hyperresponsiveness. European Cytokine Network. 2009;**20**:148-156. DOI: 10.1684/ecn.2009.0156

[104] Sousa AR, Trigg CJ, Lane SJ, Hawksworth R, Nakhosteen JA, Poston RN, et al. Effect of inhaled glucocorticoids on IL-1β and IL-1 receptor antagonist (IL- 1ra) expression in asthmatic bronchial epithelium. Thorax. 1997;**52**:407-410. DOI: 10.1136/thx.52.5.407

[105] Tanaka H, Miyazaki N, Oashi K, Teramoto S, Shiratori M, Hashimoto M, et al. IL-18 might reflect disease activity in mild and moderate asthma exacerbation. The Journal of Allergy and Clinical Immunology. 2001;**107**:331-336. DOI: 10.1067/mai.2001.112275

[106] Simpson JL, Phipps S, Baines KJ, Oreo KM, Gunawardhana L, Gibson PG. Elevated expression of the NLRP3 inflammasome in neutrophilic asthma. The European Respiratory Journal. 2014;**43**:1067-1076. DOI: 10.1183/09031936.00105013

[107] Tan HTT, Hagner S, Ruchti F, Radzikowska U, Tan G, Altunbulakli C, et al. Tight junction, mucin, and inflammasome-related molecules are differentially expressed in eosinophilic, mixed, and neutrophilic

experimental asthma in mice. Allergy Eur J Allergy Clin Immunol. 2019;**74**: 294-307. DOI: 10.1111/all.13619

[108] Duan W, Mehta AK, Magalhaes JG, Ziegler SF, Dong C, Philpott DJ, et al. Innate signals from Nod2 block respiratory tolerance and program T H2-driven allergic inflammation. The Journal of Allergy and Clinical Immunology. 2010;**126**:1284-1293. DOI: 10.1016/j.jaci.2010.09.021

[109] Żeromski J, Kaczmarek M, Boruczkowski M, Kierepa A, Kowala-Piaskowska A, Mozer-Lisewska I. Significance and role of pattern recognition receptors in malignancy. Archivum Immunologiae et Therapiae Experimentalis. 2019;**67**:133-141. DOI: 10.1007/s00005-019-00540-x

[110] Burn OK, Prasit KK, Hermans IF. Modulating the tumour microenvironment by intratumoural injection of pattern recognition receptor agonists. Cancers. 2020;**12**:1-22. DOI: 10.3390/cancers12123824

[111] Bai L, Li W, Zheng W, Xu D, Chen N, Cui J. Promising targets based on pattern recognition receptors for cancer immunotherapy. Pharmacological Research. 2020;**159**:1-13. DOI: 10.1016/j.phrs.2020.105017

[112] Nyirenda MH, Sanvito L, Darlington PJ, O'Brien K, Zhang G-X, Constantinescu CS, et al. TLR2 stimulation drives human naive and effector regulatory T cells into a Th17-like phenotype with reduced suppressive function. Journal of Immunology. 2011;**187**:2278-2290. DOI: 10.4049/jimmunol.1003715

[113] Li L, Liu X, Sanders KL, Edwards JL, Ye J, Si F, et al. TLR8-mediated metabolic control of human treg function: A mechanistic target for cancer immunotherapy. Cell Metabolism. 2019;**29**:103-123. DOI: 10.1016/j.cmet.2018.09.020

[114] Sfondrini L, Rossini A, Besusso D, Merlo A, Tagliabue E, Mènard S, et al. Antitumor activity of the TLR-5 ligand flagellin in mouse models of cancer. Journal of Immunology. 2006;**176**:6624-6630. DOI: 10.4049/jimmunol.176.11.6624

[115] Di Virgilio F. The therapeutic potential of modifying inflammasomes and NOD-like receptors. Pharmacological Reviews. 2013;**65**:872-905. DOI: 10.1124/pr.112.006171

[116] Janowski AM, Colegio OR, Hornick EE, McNiff JM, Martin MD, Badovinac VP, et al. NLRC4 suppresses melanoma tumor progression independently of inflammasome activation. The Journal of Clinical Investigation. 2016;**126**:3917-3928. DOI: 10.1172/JCI86953

[117] Lin TY, Tsai MC, Tu W, Yeh HC, Wang SC, Huang SP, et al. Role of the NLRP3 inflammasome: Insights into cancer hallmarks. Frontiers in Immunology. 2021;**11**:1-15. DOI: 10.3389/fimmu.2020.610492

[118] Chen L, Huang CF, Li YC, Deng WW, Mao L, Wu L, et al. Blockage of the NLRP3 inflammasome by MCC950 improves anti-tumor immune responses in head and neck squamous cell carcinoma. Cellular and Molecular Life Sciences. 2018;**75**:2045-2058. DOI: 10.1007/s00018-017-2720-9

[119] Tengesdal IW, Menon DR, Osborne DG, Neff CP, Powers NE, Gamboni F, et al. Targeting tumor-derived NLRP3 reduces melanoma progression by limiting MDSCs expansion. Proceedings of the National

Academy of Sciences of the United States of America. 2021;**118**:1-11. DOI: 10.1073/pnas.2000915118

[120] Yan H, Kamiya T, Suabjakyong P, Tsuji NM. Targeting C-type lectin receptors for cancer immunity. Frontiers in Immunology. 2015;**6**:1-9. DOI: 10.3389/fimmu.2015.00408

[121] Osorio F, LeibundGut-Landmann S, Lochner M, Lahl K, Sparwaser T, Eberl G, et al. DC activated via dectin-1 convert Treg into IL-17 producers. European Journal of Immunology. 2008;**38**:3274-3281. DOI: 10.1002/eji.200838950

[122] Xu F, Liu J, Liu D, Liu B, Wang M, Hu Z, et al. LSECtin expressed on melanoma cells promotes tumor progression by inhibiting antitumor T-cell responses. Cancer Research. 2014;**74**:3418-3428. DOI: 10.1158/0008-5472.CAN-13-2690

[123] Bufalieri F, Basili I, Di Marcotullio L, Infante P. Harnessing the activation of RIG-I like receptors to inhibit glioblastoma tumorigenesis. Frontiers in Molecular Neuroscience. 2021;**14**:1-9. DOI: 10.3389/fnmol.2021.710171

[124] Jiang X, Muthusamy V, Fedorova O, Kong Y, Kim DJ, Bosenberg M, et al. Intratumoral delivery of RIG-I agonist SLR14 induces robust antitumor responses. The Journal of Experimental Medicine. 2019;**216**:2854-2868. DOI: 10.1084/jem.20190801

[125] Wolf D, Fiegl H, Zeimet AG, Wieser V, Marth C, Sprung S, et al. High RIG-I expression in ovarian cancer associates with an immune-escape signature and poor clinical outcome. International Journal of Cancer. 2020;**146**. DOI: 10.1002/ijc.32818